READY SET AGILE!

BEING AN EFFECTIVE

PROJECT MANAGER

*Your Guide to Becoming
a Project Management Rock Star:
Best Practices, Methodology,
and Success Principles
for a Project Leader*

BEING AN EFFECTIVE
PROJECT
MANAGER

*Your Guide to Becoming
a Project Management Rock Star:
Best Practices, Methodology,
and Success Principles
for a Project Leader*

2020

TABLE
OF CONTENTS

GETTING SUPPORT

from the
✓ **Management**
✓ **Sponsor**
✓ **Stakeholders**

Triple Your Chances for the Project's Success

13 Tips on Getting Support

READY SET AGILE!

To download your checklist, click and visit the link:

rsagile.activehosted.com/f/15

INTRODUCTION

AS a project manager, you're used to juggling tasks, people, and resources, all while keeping the client happy. But, do you ever wish it were easier? That you and your team could do more with all those work hours you're given? Sometimes, it's a slog to get everything and everyone on the same page and all working together as one team to achieve the project goal.

Maybe you've been struggling to ensure that projects are completed on time and budget. You may think you've estimated the resource needs correctly, and then there's suddenly an unexpected and unforeseen development that ends up destroying the timeline. As of the time of writing this book (June 2020), we're still in the middle of the coronavirus epidemic. No one saw this coming, and it has upset the business plans for quite a few project managers.

No one can reasonably blame you for not seeing the COVID-19 virus coming beforehand. However, do you find that you're often held accountable for issues that are out of your control? As the project manager (PM), *you* are the one who's responsible for the completion of the project as planned. Despite this, project teams or stakeholders may sometimes seem to take it too far and blame you for things that genuinely were not in your control.

Stakeholders are known for pressuring their project managers and teams. Other issues may have come up, and they'd like to shift some of the budget they agreed to invest in the project somewhere else. And they may want you to finish ahead of time, despite signing off on the months of work that you estimated the project would take. Maybe they want to add some additional features to the scope without increasing the budgets for time and money appropriately.

Staying organized in terms of time can also be difficult for a PM. Once the schedule starts to slip, it's not always easy to get it back where it needs to be. You're also responsible for organizing the work, and some tasks must be completed before the next one begins, which may cause headaches when your resources aren't available as expected. How can those tasks be completed effectively, so the project moves on to the next phase in time?

Did any of these issues resonate with you? If so, you've come to the right place! This book is intended for project managers to elevate their game and create solutions for their problems. You'll learn key elements of being a project manager, along with step-by-step methods for improving in the areas where you need help. Some of these are based on leadership fundamentals, like communication and collaboration; others are focused on the technical skills required for the project's management.

This book is not necessarily about the technical skills that your team members will bring along with them, though you will see me promoting the idea of learning the basics of a project team's tasks. We will be discussing more about the skills specific to management and not the underlying functional knowledge that your team players will bring.

In addition to benefiting you as a PM, how does knowing how to be an effective project manager help others? Being effective is a huge part of making the project successful. You're more likely to bring projects in on time and at the expected costs, which makes you more likely to complete the project successfully and satisfy the stakeholders. Your effectiveness means that their expectations have been met, and they're pleased with the results you delivered.

Your absorption of this material is also to the advantage of your project team(s). When you're doing your job well, they're freed up to do their best work as well. If the environment you provide them is healthy, their morale will be higher, and they will be more productive and willing to go the extra mile for you.

When you, as the PM, understand where you should prioritize your work, look out for potential threats to project completion that are looming on the horizon, and how best to communicate with all those involved on the project. If you do so, everyone will be happier with the results.

Being more effective on your projects has personal benefits for you as a PM. Project managers who can bring in projects consistently and according to schedule and budget are rock stars in the project management world. Clients will want to hire you because they will notice your successful track record. Besides, you'll enjoy your work more since the daily tasks won't seem so difficult. Once you can get the project onto a consistent schedule, and all the team members and stakeholders are moving in the same direction, projects won't be as difficult to complete. You can then enjoy yourself and still get an excellent job done.

Effective PMs can remove a lot of pressure. You'll still probably run into your share of stakeholders who just want to squeeze what they believe is a quick upgrade into your meticulously constructed framework. However, once everyone is on your side, there will be much less external pressure, and clients will be more willing to give you the benefit of the doubt.

The more projects you work on, the more you can boost your career. New projects mean a wealth of lessons and experiences that will multiply your level of expertise exponentially. I'll show you how to spot these ideas and leverage them to your own benefit.

Finally, you'll develop winning teams that are rock stars in project management. They'll want to work with you in the future because an effective PM leads their team in a way that makes it a joy to work on projects. They will also, naturally, give their best in return. They'll respect you, not only for your success, but also for understanding their jobs and how you let them work most effectively. We know that people are happier at work when they have some control over it, so to the extent you can empower team members to accomplish their jobs, the better it will be for their morale.

At this point, you're probably wondering how I can be so sure about the results of effective project managers! I've participated in projects and been a PM myself in many different project management positions. I've collaborated on a variety of projects, from small start-ups to large, cross-divisional ones with huge budgets and a timeline of several years. In my career, I've seen projects succeed, and I've also seen them fail. My goal is to help aspiring PMs learn how to be effective quickly without spending years learning it on the ground as I did. Throughout all my years of project management, I've learned vital lessons on how to be an effective project manager, and I want to share this with as many PMs as I possibly can.

By using my experience and knowledge, you'll learn the key elements of project management and how to improve in the areas where you don't feel as confident. You'll understand how to win over the sponsors of your project and have them cheering you and the team on while removing any impediments that they have influence over. You'll have the keys to executing the plan, technical skills, and knowledge of project management as a business. In addition, you'll learn the basics of the most important project management (and life) skill: continuous improvement.

Many other PMs, just like you, have used these principles to build better lives for themselves. Being an effective project manager often leads to personal improvement as well. After understanding and implementing the material that I've gathered for you here in this one place, they've gone on to lead successful projects. Such is all just from the simple steps that have been outlined in this book.

Many PMs weren't sure they were really cut out to be rock stars, but after becoming more effective, they went from zero to hero pretty quickly. For me, some of these transformations were a little short of miraculous. Not only did the lessons benefit them, but their entire team and other stakeholders reaped the benefits too.

Why wait to have a career that's fun again? One where you don't have to work harder, but just a little smarter to get to where you want to go? If you're already a project manager and you feel like you're struggling through your workdays, don't hold off on reading this book. Start making your work life a little easier by increasing your effectiveness today.

Effective project management at work is great, but it also spills over into your personal life. For one thing, you're not trying to roll the ball uphill, so you have more time for family and friends. When you get more organized in one aspect of your life, you will tend to take that skill into all the others too. As you improve your communication with stakeholders and team members, you will become more effective at home as well. You'll initiate and complete personal projects to good results, and your ability to hire and deal with the right specialists to make your projects successful will increase.

You may be wondering—what exactly does *effective* mean when it comes to project management? It's about achieving common goals with given resources in terms of money and time. It also entitles becoming a skilled prioritizer who can focus on the work and deliver and creating value for the business through the successful completion of the project. It means you do the work faster and smarter, take your projects further, and—most importantly—enjoying the work that you do.

Sound good? Let's get effective!

CHAPTER

1

THE BEGINNING
OF AN EFFECTIVE
PROJECT
MANAGER

THIS book is for anyone who wants to be the most effective project manager (PM) they can be. That is whether you're already a project manager looking to level up, or you're still aspiring to become one. Here, you'll discover what you need to know, so you can hit the ground running right after reading and practicing the tips in this book.

What, then, makes a good project manager? It's not the title, position, education, nor experience, although these elements can help tremendously. It's something within you. Let's discover this something together.

What is the Role of a Project Manager?

Roughly speaking, a project is a group of tasks that must be performed to achieve a specific outcome, whether that be something new or improved. This can be a project that you fulfill at home, such as renovating your house, or at your work. Projects can be found in all industries, including software, manufacturing, and financial services. Any department within a company can manage a project to improve the current process or deliver a new product. Project complexity can range from fairly simple with few tasks to extremely complex, and they could require just one team or significantly more.

No matter what type of project you're considering, there has to be someone to oversee the entire process, and that person is the project manager. Titles may vary in the agile project management world; for example, a Six Sigma team might be led by a Black Belt or Master Black Belt, and a Scrum team could be led by a Scrum master, (see my previous book *Become an Agile Project Manager* for more details on these kinds of positions).

In the world of traditional (or "waterfall") project management, the PM would keep a tight hold on the reins. They typically negotiate with the client representatives, develop the budget, hire the team, organize the tasks, and create the schedule. It's a top-down style of management that works for projects that have specific delivery dates and where one task cannot be started before the previous one is completed.

Agile project management teams, by contrast, are self-organizing and all the team members are involved in client communication, budgeting, and figuring out which tasks should be performed next. Agile teams have a system in place (known as the artifacts) that permit them to keep track of what is required, has been accomplished, and needs to occur next. These projects are run in short sprints or iterations, so the teams would manage their own schedules. Agile is better for projects that may require quick adaptation to changing circumstances, such as software development.

Whichever project management type you're using, there's a need for someone to keep an eye on the entirety of the project. There are several characteristics and skills that a PM needs to be effective for any project, which are a combination of technical and leadership skills.

- **Prioritization**

  There are always enough tasks to fill up 25 hours of your 24-hour day. Not all of them are important or necessary, and some you may be able to ignore altogether. To be a successful PM, you need to determine what's most important for your team to accomplish in the next time period (day, week, etc.). Once that is done, move to the next, then the next. Besides, you'll probably need to help your team prioritize their own tasks.

  Do the right thing at the right time.

- **Organization**

  The role of a project manager involves juggling a lot of different tasks and concepts, with plenty of minutiae and distractions. If you're not organized, you'll be snowed under the high volume of work.

  Two main ideas contribute to being well-organized—one is having a tidy space, represented by Marie Kondo (n.d.).[1] Your work surfaces should be neat when you arrive, so you're not distracted by the mess. Everything that you need for your work needs a home, and everything that's not necessary should be stored for when you need it; otherwise, it should be eliminated if you have no need for it at all.

The other thing you need to organize is your time. Prioritization helps you stay on top of what's urgent and important, and a successful project manager will ensure that their time is productive.

Teamwork

As the PM, it's your responsibility to make sure that the team is working together. Otherwise, you won't be able to make your deadlines, and the project itself could fail. You'll need to mediate conflicts when they arise, remembering that healthy disagreement is positive for the team.

You'll be a player-coach on this one, whether you're working on a traditional or agile project. You're building a team, not just with your own workers, but also with clients, client representatives, vendors, and other stakeholders. As you work with others, you will show your team members more about how it's done.

Communication and diplomacy

Though agile places more emphasis on communication than waterfall does, in both situations being able to connect with other people is a requirement. Your team members may only be talking to each other and the PM, but you have several audiences to check in with. A key function for a PM is to provide understandable and actionable updates to stakeholders, where necessary.

You'll likely be dealing with people from a variety of backgrounds and educational experiences, and all of them need to understand you clearly. In addition, communication can't be offensive or harmful to anyone; it should build relationships rather than damage them.

Hence the importance of diplomacy as a component! Later in this chapter, we'll discuss emotional intelligence and why it's necessary—but for now, just understand that being able to deliver bad news without severing the relationship is key. Projects may go over schedule on time or budget, the team may misunderstand the client and make a mistake, or some other issue rears its ugly head in the middle of the project. Stakeholders need to be informed so they don't become excessively upset or pull the plug.

● See the big picture

Ideally, the PM would understand all the different tasks that their team performs. They may not be an expert in any of them, but they would have a working knowledge of what each person specializes in. They're not the ones doing the work necessarily, but they are overseeing it. They're the ones who must see how all the pieces in the puzzle fit together, even if each individual team member only knows what they themselves are working on.

The PM would understand how all the tasks come together to create the deliverable, and what needs to happen before the next task can commence. This also includes what the team and stakeholders need to know about the process, and how to integrate their input into the project. They should perform the research necessary to pull everything together in a puzzle that evolves dynamically.

● Fresh perspective

One thing that machines and robots can't do for us is to give us a new perspective on a problem. They can churn through data and create charts and graphs to make the statistics more understandable, but they can't turn the problem on its head to see if they should shake loose some different ideas that way.

The more you (and your team) can be creative, the more viewpoints you will have on the situation, and the more likely you will be able to develop a good solution to the problem.

• Critical thinking

Unfortunately, most of the time, you won't have complete insight into decision making. You may be able to see your options, but you won't be able to project them out into the future with a full guarantee that trying X will work, or that Z will work better. You may not have clarity on all your choices, or you may not have the resources for your first and best option.

Successful PMs can think through what they know—to paraphrase Donald Rumsfeld, former Secretary of Defense for the United States, the known knowns and the known unknowns to reach a conclusion. They can also analyze the situation when the unknown unknowns show up.

• Leadership

As a PM you're leading the team, and a strong PM inspires and motivates the team. Modern leadership uses emotional intelligence skills to communicate and build relationships and connections. This is particularly with the team, but with others, such as stakeholders, as well.

You would set the vision; then, you would coach the team to achieve it. You're personally responsible for the success or failure of the project. In other words, you're not a manager who can blameshift up or outwards—you're the leader who goes down with the ship, or reaps the rewards of a well-done project that reaches the objectives.

Project Leader vs. Project Manager

Although the title is *project manager*, a great PM is still a *project leader*. We alluded to it in the previous section, but what does it really mean to lead a project as opposed to merely managing it?

Very briefly, the difference comes down to strategy compared to tactics. A leader is focused on the strategy: the big picture and what needs to be done (at a bird's-eye view) to accomplish the objective. They empower and inspire their team members to help them reach that vision and get the job done.

By contrast, a manager is on the tactical side: figuring out what needs to be done in what order, and who should do what. In this book, we'll be discussing the role of a manager rather than a leader, which will be covered in a different book. However, these two roles go hand in hand, and a successful PM has to master both.

In agile, the difference between project management and project leadership is more apparent. Because the teams are self-organizing, they would decide what needs to be done next and in what order. They're usually cross-functional, so they can decide amongst themselves who is responsible for which task and when. The project manager sets the vision and strategy for the project.

In waterfall project management, the PM may act in both roles. At the very beginning, they would decide strategically with the client on the deliverable. They would provide the vision for how the project would work, then persuade the stakeholders to buy into that vision. After that, it's manager mode time as they determine schedules, budgets, and who is responsible for what and when. In this type of project, the entire process is mapped out at the beginning, soup to nuts, which is mostly the PM's role, though they can consult with the team members who have expertise in individual sections.

Leaders need to demonstrate three types of mental competence. A manager may be able to get away with two of the three, but a true leader shows evidence of all of them.

1. IQ and cognitive skills

This is measured by IQ tests and mostly involves problem-solving and analytical skills. It's fixed at birth. Although you can acquire more knowledge, your IQ is relatively static over your lifetime. On the other hand, cognitive skills such as paying attention, retaining memories, and the ability to both plan for and execute the steps necessary to achieve a goal can be improved when you spend time learning and practicing them.

2. Emotional intelligence

Also known as EQ, this type of intelligence is about how well you can build and maintain relationships. People with high EQ are aware of their own emotions and can regulate them. They also help others to manage their emotions.

Unlike IQ, you can increase your EQ merely by applying yourself. Those who work for supervisors with high levels of emotional intelligence are 400% less likely to leave the job (Leadem, 2018).[2] You can see why this aspect is essential for leadership!

3. Managerial intelligence

This type of intelligence (MQ) measures your knowledge of skills and the qualities necessary to manage others (Zilicus, 2014).[3] It's about managing and controlling resources (material and human) and achieving the goal. You can learn this skill as well to improve over time.

A study of these different types of intelligence showed that EQ is most important to the success of a project, contributing 36%; IQ 27%; and managerial intelligence 16% (Dulewicz, et al, 2016).[4] You can see that emotional intelligence is critical for leaders, but there are additional issues that come up for a project manager to be a project leader (for the rest of the book, understand that an effective project manager is also a project leader). One of the aspects of the project that must be managed are the constraints on the project itself.

Triple Constraint Triangle

Don't worry—good knowledge of geometry isn't necessarily required for project managers, effective or otherwise. The Triple Constraint triangle is a way to visualize the boundaries on the project. Those three are the points of the triangle: time, scope, and budget, with quality being the central theme of the triangle. You might also see more recent adaptations that describe constraints as a square, with quality being the fourth point and expectations as the theme of the square.

When you're thinking about your project, you will need to manage all these limitations.

- **Cost**

   Your clients don't have infinite amounts of money to throw at the project you're managing and need to see a good return on their investment. They would allocate a certain amount of money, and it's your job to deliver within that framework. Clients really dislike it when you ask for more money; therefore, in general, the budget you draw up in the beginning should be the maximum you should expect.

- **Time**

   Clients need their projects delivered for a specific deadline. For example, if you're managing a project to develop a new toy, the firm will want the completed project in advance of the end-of-year holiday season to ramp up production, marketing, and sales.

   PMs who deliver projects routinely on time are extremely valuable to their clients. Clients want to know they'll get their result when they were expecting to achieve it. Being able to do so consistently makes a project manager very attractive.

Scope

Whatever you agreed to deliver must be satisfied by the completion of the project. That also means that both you and the client need to be clear on the scope, which is often where misunderstandings occur.

For example, if you're working on a software development project, the client may think they've contracted for a fully functioning, tested, and bug-free app. However, the PM may have understood that they're only responsible for a beta version. Ensure that you ask the right questions and the scope is spelled out clearly in the documentation.

Quality

Clients want deliverables that will stand up to usage. However, some clients may be OK with a version that's closer to beta than full release, in the parlance of software development.

The project manager must understand the expectations that the clients and stakeholders have in terms of quality, and they're responsible for delivering it. Clients want results—not tick marks on a checklist.

Expectations

Like scope, expectations are often misunderstood. As a PM, you'll need to ask probing questions to be clear on what it is they think they will receive at the end of the process.

In an ideal world, of course, there are no errors or wrenches thrown into the works. Everyone who was hired to work on the project stays until the bitter end and manages their work so the project can stay exactly within the budget and time constraints. All team members work to their highest quality, and the deliverable should be of the finest quality possible.

"You can have it fast, cheap, or good—pick two." —Anonymous

In the real world where you'll be operating, the project will not flow smoothly from beginning to end and without bumps in the road or obstacles in the way. The Triple Constraint is about how you shift from one point to another to achieve the best compromise possible when you can't make all three.

For example, once the project starts going over schedule, it's extremely difficult, if not impossible, to get that train back on track and make your deadline. You will have some decisions to make, and, as a PM, you'll be the one held accountable for them. You could allow the project to simply run over the deadline, but the client may not accept this. You could hire more people or other resources to make up the time, but that will put you over budget, unless you gave yourself some wiggle room for exactly these kinds of situations. On the other hand, you could pull back a little on the scope to finish on time, which will decrease the quality.

Often, the client will let you know what's most important to them. They might stress to you that the project needs to be completed within the budget they provided to you and not a penny more. In that case, you may allow the project to continue with fewer resources past the deadline.

There could also be a drop-dead delivery date put in place, in which case, you might propose to hire more resources. If they've given you this kind of guidance, balancing the three constraints may be a little easier because you know what you have to hold steady on and can then figure out how to massage the other two issues to achieve the goal.

When you don't know what's most important, you may have to play with your problem-solving a little bit more. For example, to avoid a complete trainwreck, you may choose to go a little over budget and time and ease up on the scope and quality a little as well. Depending on the project and the client, there may or may not be a clear solution for how to balance these three constraints on your project.

You'll need to communicate your decisions to the team and stakeholders. Depending on the style of project management you're working with (waterfall or agile), both the stakeholders and team may have to be involved in the decision. Ultimately, it's your responsibility and you're the one held accountable, but your team members may come up with a solution you hadn't considered. The stakeholder may be able to squeeze a bit more out of the budget if they understand why you're asking for the additional resources, and they might have some expertise that they can bring to solving the problem as well.

Given that no one—or at least not everyone—will be pleased with the news, your communication skills as a PM will become extremely important in these situations! You need to be direct and clear and not passive and wishy-washy because you won't like delivering bad news. At the same time, everyone must be clear that you respect their opinions and are willing to entertain other ideas and solutions. Involving the client representatives in the decisions themselves will help you get buy-in from the clients. This will assist enormously when you're asking for extensions of time or extra resources. People dislike being told what to do, so as much as they can feel in control, or at least have a voice in the process, the easier the constraint balancing will be for you.

You can also see that the more respect your team has for you, the easier the balancing will be. We'll get into more on how you can inspire and motivate your team members to respect you later in the book.

Jack of All Trades

When you know fundamentally what each of your project workers is doing daily, you will be better able to inspire the respect of the people on your team.

Have you ever worked for a boss whom you felt didn't understand your job? Were you frustrated that they couldn't understand how long certain tasks took because they had no concept of what actually went on during the completion of the task? Did you and your coworkers ever make fun of them behind their backs for being so clueless, or make remarks about how they fulfilled the Peter Principle?

"People tend to rise to the level of their incompetence." –
Dr. Laurence Peter (Wikipedia, 2020)[5]

This suggests that, as people get promoted, they become less effective because doing well in one job doesn't guarantee the same in another.

You don't necessarily have to master each of the skills that your team members have or be able to do their jobs at the drop of a hat. Since most of the time you're also hiring team members, you will need to be able to judge their level of performance. If you have no idea what they're doing and what's involved in their work, you will have no way of understanding who's better at the task than another.

As a project manager, you're the one who fits the puzzle together to complete the project. To do that, you'll need to know the size and shape of the different pieces. When you're managing a software project, for example, not understanding basic programming and data structures means you will have no idea what the shape of the pieces are, thus you won't be able to fit them together logically.

By contrast, when you understand the fundamentals of the tasks that will be performed, you'll also have a clue about how long they take, the kind of skills a person would need to fulfill them, and the resources they would need. Such allows you to view all the pieces clearly and put the puzzle together in a way that best fits the picture you've agreed to deliver to the client. Complete with timelines for each task and team member, and provide allowable expenses for each.

You're Part of the Puzzle Too

Don't forget yourself when you're figuring out how all the pieces work together. Whether your project is waterfall or agile, you'll be encouraging your team members, supporting them in work, keeping an eye on the details, and reminding everyone of the vision that you're trying to achieve. You will need to be a leader—not just a manager of people. The qualities that you model can make or break a successful project, and you'll inspire everyone working on the project to follow your lead.

If you're not very effective and waste a lot of time and energy, so will everyone else. If you promote playing the blame game and avoiding accountability, you'll find that no one on the team will ever be responsible for anything that went wrong. Client representatives may even throw you under the bus when they're talking to their superiors about what went wrong on the project. If you get defensive when others offer criticism, no one on the team will offer up suggestions for improvement. If you can't communicate with people at all levels of the organization, and from all backgrounds and experiences, you'll end up at an impasse when you can't get your message across to the people who need to hear it.

The good news is that you can learn to be an effective PM and avoid all those mistakes. In the rest of this book, you'll discover all the pieces you need so you can deliver the best results. It's not just about specific roles that you would take on in the project, though they play a role and we'll discuss them in more detail; it's also about leadership skills and how you can improve yours.

The paradigm has shifted away from command-and-control style leadership, in which the PM is infallible and up on a pedestal. Waterfall style project management is more about command and control, and, as such, it's fallen out of favor. It has a rigid structure and planning is done almost entirely at the start of the process, which allows for less freedom and flexibility.

Agile is more successful, and more companies have begun to adopt this faster style that allows them to go to market sooner and reap their return on investment quicker. Similarly, being a project manager is more about encouraging and coaching the team to achieve their goal instead of providing carrot-and-stick incentives.

You'll discover everything you need to know about being an effective project manager and put all the puzzle pieces together to deliver solid and successful projects when reading through this book.

Key Takeaways

Effective project managers understand how to put all the pieces of a project puzzle together to deliver rewarding results and experience for the clients, the team, and themselves.

- Certain qualities, such as good communication, teamwork, and being well-organized, are essential for the effective project manager.

- Project managers who want to be effective and successful need to lead a project strategically; not just manage it in terms of tactics.

- PMs must balance the three constraints of a project—scope, time-frame, and cost—which is visualized by the Triple Constraint triangle.

- A trait that helps project managers balance the triangle is to be a jack of all trades, in which they would have a fundamental understanding of all the tasks that need to be accomplished.

- The project manager is the one who fits pieces of the puzzle together.

- Anyone can learn how to be a more effective PM.

In the next chapter, you will learn about the key components for a successful project.

---

[1] https://**shop.konmari.com**/pages/about

[2] https://**www.entrepreneur.com**/article/318187

[3] http://**zilicus.com**/Resources/blog-2014/Project-Leadership-Or-Project-Management-Becoming-Effective-Project-Leader-Part-ii.html

[4] http://**doi.org**/10.4236/ajibm.2016.64043

[5] https://**en.wikipedia.org**/wiki/Peter_principle

CHAPTER 2

KEY PROJECT

COMPONENTS

AND THEIR

ASSOCIATED

SKILLS

AREN'T all the aspects of a project important? Well, yes and no. There are a few parts of a project that have a much larger effect on the success—or lack thereof—compared to others. This is true whether you're managing an agile project or a traditional one. If you can get these significant components right, you'll set yourself up well for success. Projects tend to fail because the following aspects weren't properly planned or communicated.

Scope

In the first chapter, you learned that it's common for the scope to be misunderstood between the project manager and the client. However, just because it's common doesn't mean that PMs shouldn't avoid this confusion whenever possible.

One of the major malfunctions in defining the features of the product or process to be delivered is known as "scope creep." That is, the project management team or client adds on to the original features of the project. Once the scope is broader than what was originally promised, the timeline or the budget, or both, will be put under sudden pressure. Scope creep can affect the quality as well. The higher number of elements added, the more the entire project may degrade in quality.

It's important to know that, often, scope creep doesn't happen all at once. Sometimes, the client may ask for something extra at first, and then they seem to forget it wasn't included in the original scope, timeline, and budget. Clients ask for additional features mostly because they may not have been sure what they want exactly in the beginning. If the client is unclear at first, then the team may not gather enough information for the requirements analysis. Without truly understanding what the client wants, it becomes easy to continue adding new details as the client asks for them.

Ultimately, the PM is responsible for ensuring that scope creep doesn't happen on their watch! Making sure the requirements analysis is sufficient is one task that they oversee. There are also many reasons not necessarily related to the client that can be responsible for scope creep, depending on the circumstances.

• A project that has never been done before

If it's a new project, or the PM and team are new to the project type, they may not realize the complexity of it initially. If no one on the team has that much experience in the subject, there will be no one to ask and no use cases to examine from previous projects to apply as a reference.

When that's the case, it's best if the PM expects cost and time overruns and budgets accordingly. It's highly unlikely that the project will be less work than you expected. Assume it will be more and ensure that you have some contingency planning in place to deal with extra complexity as it arises.

If those types of projects have been successful (or not) in the past, a smart PM will study those use cases to decide the timeline and budget required to manage it successfully.

Analysis paralysis

Studying and researching, in so many ways, is so much easier than actually entering production! You don't have to take responsibility for anything going wrong while you're researching the problem. With all this information overload in the world, it's incredibly easy to go down the rabbit hole and never get anything done. After all, you could spend an evening easily reading all the articles about scope creep on the Internet! You may only get started when the deadline is looming and you're starting to panic.

Analysis paralysis is especially common among perfectionists. They find it difficult when they start on production that does not yield perfect results at first, and they prefer to continue analyzing instead of working. Once they *do* get started, they may be reluctant to ship or send their piece of work for testing because it isn't perfect yet.

"Done is better than perfect." —Anonymous

Workers downstream from the perfectionist often get very frustrated because they don't have what they need to do their own jobs until the perfectionist finishes theirs. It is much better for the work to be released on time, so the rest of the team can perform their work. If you wait for perfection, you may never have any product at all. There's always a tweak that could be made or a little touch-up, but the product needs to be released at some point.

No control over changes

If the PM doesn't take the reins of making changes to the scope, the project is much more likely to be a victim of scope creep. Instead, the PM should include upfront documentation that any scope changes must go through an approval process. When the client can't drop in on the team members and request a change, they'll give it more thought. The PM can also specify in the documentation that extra time and costs for the change must be approved, so the client is aware of the consequences of their request.

Overcommitment by the PM

As discussed in the previous chapter, although, as a project manager, you may have a basic knowledge of the various tasks and skill sets required for the project, you may not know enough about the learning domain to estimate its scope accurately.

Never make a commitment to the client regarding a function you're not well experienced in without consulting your team members, who are the experts in their domain. Only after they've given you the approval on a reasonable timeframe can you go to the client and make that agreement.

Not getting users involved early

This issue is particularly important when it comes to software development. The client representative, if they're in management, may not have understood what exactly the users need. In agile project management, users are involved as part of the process, but that isn't often the case in waterfall.

They need to be involved from the *beginning*—not just toward the end when the team delivers the product. If they're brought into the process late and the project isn't providing what they need, the scope must be changed (with its deadlines and expenses) to ensure that the team delivers something that is valuable to the users.

The users provide valuable feedback to the team on what's working and what isn't. When done early enough in the project, changes can be made that won't necessarily disrupt the timeframe or budget.

Gold plating

Sometimes, the team members will add to the project believing that they are creating more value, which is known as gold plating. It's common in software development, but not exclusive to that industry. The changes they make don't necessarily add any value to the project, especially if they're not communicating regularly with the users. The gold plating eats up time and money without guaranteeing client satisfaction.

It's not relevant whether the team thinks it has created value; the key metric is whether the client believes they did.

Value

The project is initiated to create value for the business or client. Keeping that at the forefront of their minds in terms of the goals and daily work will help the project management team supervise the scope.

Creating a deliverable of value is a matter of pride and accomplishment for the team, and working toward a valuable goal helps team members stay motivated and excited for the work. It also assists them in working together as a team. Petty or interpersonal differences are set aside when everyone is seeking to achieve the same objective.

The team's contributions have value, which must be recognized by the project manager if no one else. Through their daily work, they would contribute to the value of the project. A good PM who acknowledges team member input as valuable can reduce the extras that they may otherwise feel they need to add.

A project manager is also part of the team and should recognize their contributions to value creation. By maintaining a clear vision of what the project is intended to achieve, they can lead the team to success. Staying on the scope and creating the value as mapped out by the client leads to happy clients and satisfied teams.

Thinking Ahead

All projects require a significant amount of planning, even agile projects, in which the project manager expects there will be changes and adjustments throughout the process. With any style of project management, the PM must engage in planning right from the beginning, once they understand entirely what the scope of the project is. There are several different types of skills in this area that a good PM would demonstrate in their work. For most of them, the more projects you can manage, the more you will improve.

- **Time planning *(aka scheduling)***

   Clients want their project completed by the time that you agreed with upfront. Missing delivery dates makes you a PM that others won't be willing to hire. Being able to estimate how long the various tasks will take is key to agreeing to a doable date.

   If something happens and you realize the date will be missed, you need to be able to communicate such to your client and work toward a new delivery date that you won't miss.

   Early isn't necessarily a positive thing either. Some clients operate on a just-in-time basis, and they won't appreciate having the product taking up space if they can't use it right away. There may be times and projects where the client will appreciate an earlier delivery, but make sure that's the case first if it seems you overestimated the time required.

Lean heavily on your team members with expertise in projects in their domain for solid ideas of each task's timeframe, especially if you haven't done many of these particular projects before. You can also try to find some use cases from similar projects to see what their timeframes looked like.

Cost estimation

As you know, clients hate it when you come back to them with your hand out asking for more money. Get the cost right as *best* you can, and allow yourself some contingency funds *if* you can.

As with time management, it's a good idea to consult with your subject matter experts before bringing a budget to the client if it's a type of work or function that you're not as familiar with.

Resource planning

Knowing how much material and people you will need at the beginning of the project is essential. Rather than continually requesting new items as you go, you can take advantage of the firm's purchasing power in bulk to reduce the cost as much as possible.

People tend to be expensive relative to other resources, so you're better off having enough money in the budget to hire as many people at the outset as you'll need. This is somewhat less of an issue with agile because the team members are cross-functional and can fill in at a variety of tasks. However, in waterfall, you may require some subject matter experts and need to hire people to handle those specific tasks.

Again, it's helpful to consult with others before including any final numbers in the plan you bring to the client. You may have people on your team who can help you make these kinds of estimates, and you can also check the literature to see how similar projects behaved.

● Critical thinking

Successful projects require project managers who can make the right decisions, even under conditions of uncertainty. It's being able to analyze the choices on offer by considering the advantages and disadvantages, along with projecting the likely consequences of each option. Critical thinking allows you to consider the long-term costs and benefits. Then, you can make a confident decision based on the available information, which is usually not all the information you'd like to have before making a choice.

Although you can consult on possible solutions to a problem, ultimately, you're on your own as far as critical thinking goes. You can't outsource it to more experienced team members or study use cases, although you will get better over time with continual practice.

One technique that many critical thinkers use is to be contrarian. Being contrarian is considering the opposite viewpoint to the one you currently have and the benefits of that viewpoint. Using this technique can help you broaden your choices and think more creatively about the problem.

● Strategic thinking

As the leader of the project, you're the one who needs to think about the long-term process of the project. Your team members will be focusing on the daily, tactical aspects of it: who does the next piece of work and how. You'll need to manage the daily tasks to ensure they fit in with the long-range vision. Team members have some of the pieces of the puzzle, and you need to make sure they all fit together in a coherent picture.

You're also the one who must scan the horizon for possible threats and opportunities. For example, a supply chain disruption overseas could affect the delivery of the resources you've requested. The project manager is responsible for figuring out how to deal with or mitigate the obstacles that inevitably show up during a project.

Like critical thinking, it's a skill that you can't outsource to anyone else. Some barriers to the project are foreseeable, and you can deal with them ahead of time by putting a contingency plan in place. However, some of them will just show up randomly for you to handle.

Being able to listen actively to others' viewpoints will help you improve, as will learning to ask the right questions. That's mostly a matter of practice, as is being able to think through the potential consequences.

One way to practice is to consider not just the first-order consequences (or immediate results), but also about the second order, third order, or results of the results. You can do this for simple decisions in your personal life for practice.

For example, maybe you're considering getting a pet. You know you'll need to prepare your home and pick up bedding, food, and other items. What happens as a result of selecting the bedding? You'll need to find a place for it where the pet will be comfortable. For this example, let's say the only place you can think of is your bedroom. Your new pet might cry at night or make noise because they're not used to their new home. What will you do to mitigate that?

Managing Risk

Have you ever seen a group of meerkats at the zoo or on TV? Most of the animals are scurrying around on the ground, eating, drinking, or hanging out. While they're doing that, there's one that stands up on its hind legs, using its tail for balance, and acting as a lookout. It watches for predators or other threats, which can come by air in the shape of hawks, or on the ground like snakes and jackals.

That's you as a project manager: the lookout meerkat. Your team members are on the ground doing the work and attending to the tasks necessary for project completion. On the other hand, you need to be up on your hind legs and scanning the horizon for threats, no matter where they may come from.

At the beginning of the project, you should be able to list the known risks. Such would provide you with the ability to make contingency plans; so, if the known obstacles emerge, you can shift to your back-up plan or engage the mitigation strategy that you chose. It's important to do this so you can have the budget and resources on hand to deal with problems that you and the team believe are likely to occur.

As with planning skills, you'll improve your risk management as you lead more projects. You can also consult with your team and check the literature for similar projects to see what kinds of risks they faced and how they handled it. This is especially helpful when going into a project, but still, you may not have time to read previous use cases when the unexpected occurs suddenly.

Unlike the lookout meerkat, you're accountable for ensuring that the unexpected wrench in the works doesn't affect your team. If a meerkat threat appears, they will head for their burrows. You need to keep the team working as much as you can; therefore, in addition to mitigating the potential issues known at the start of the project, you'll need to manage the ones you didn't see coming.

You can practice identifying risks or threats, and assigning them probabilities. Some risks have a higher likelihood, but they will have little effect on the project, whereas others may have a lower probability but may stop the project in its tracks. Practice being creative in handling these risks too; you may just find a way to avoid the risk entirely rather than putting your energy into mitigating it.

Be Proactive

As a project manager, you must be a self-starter. You're the one who's large and in charge, and you can't wait for people to tell you what to do and when to do it. You'll often be the person telling others what to do and when! If something isn't working quite the way it should, you have the ultimate responsibility to figure out what's happening and fix it. The bottleneck may be at the client level or in the team. Either way, you'll need to spot it and unclog it so the work can continue.

This is also an indication of maturity. There will be times when you don't feel motivated and you don't want to go to work or tackle a problematic task. However, you still have to do it anyway, without complaint or delay. Project managers must be mature enough to handle roadblocks and a lack of motivation for themselves, not to mention the entire team.

Being proactive is another skill that doesn't lend itself to consulting others. It's a trait that project managers must possess to achieve their objectives. Fortunately, it is something that you can get better at if you're not as self-starting as you'd like to be.

Make sure your focus is on what you can control. If you're constantly worrying about things that you can't influence, you'll start to get frustrated and get burned out pretty quickly. Look only at the issues that you have at least some influence over and let the others go. Practice your prioritization skills, so you ensure that you take care of the issues that affect the project the most first. You won't always know that what you're doing will be successful, but as long as you've thought through the decisions and likely consequences, take action without worrying about the results.

Key Takeaways

A handful of concepts are critical to the success of the project, and you'll need to acquire the associated skills to achieve your goals.

- Scope creep is common and, as a project manager, you need to do everything in your power to avoid it.

- Value creation and commitment are essential for the project and help the team stay on the scope.

- Effective project managers need to master a variety of planning skills.

- They also must be strong risk managers, both for known issues at the outset of the project and of unforeseen obstacles.

- Being proactive is not only a sign of the required maturity but is also necessary for the successful completion of a project.

In the next chapter, you will learn about taking charge of competency and what it really means in the context of project management.

CHAPTER 3

WHAT IS COMPETENCY IN PROJECT MANAGEMENT REALLY ABOUT?

IN an age of specialized skills, many people misunderstand the core competencies for being a leader. Project managers do not need to be experts in every technical task their team completes for the project. They don't necessarily need to be software developers, toy manufacturers, or quality control specialists. It's better for them to have an understanding of these skill sets, but they don't need to master them. The skills that project managers need expertise in are the fundamental skills of a leader.

The Expert Role in Project Management

There are several ways that experts would get involved in the project. Often those who believe themselves to be experts really aren't, which the true experts find greatly annoying! As noted above, the PM isn't expected to be a subject matter expert in the tasks that the team carries out. You definitely want to have team members who've mastered software development when you're working on a software project, but you don't need to be an app master yourself.

Find out what the client is an expert in, which can be helpful on the project. Usually, one area of the client's expertise is in their own company. They've mastered (hopefully) who does what and who can give the necessary permissions to obtain more resources or whatever's at issue. Some clients do know what they're talking about and can be helpful when involved in the project. Others, not so much. When you do have clients who aren't experts trying to be more involved in the project, only to slow things down, it's your responsibility to keep them out of your team's hair so the latter can go about their tasks efficiently. Still, you need to value the client's ideas, even if they're not ideal because they're providing you with a project to manage.

In addition, you need to remember that, even if you're not a master at the underlying task, you are still an expert regarding the project management process. Your client hired you for a reason, and that's to ensure that the project is successful. Even when you're new to the field, you probably know more about project management than the client, so be confident in your communications with them.

Competence as a Leader

You need to be proficient in your project management skills to be seen as competent. Note that I said project management—not functional tasks. As noted earlier in the book, understanding the basics of what your team members are doing is extremely helpful. However, you're not the one who has to be proficient at coding when you're leading a software development team, for example. Those are your functional experts.

The skills in which you need to be proficient are from the strategic perspective, not the daily or weekly tactics. Your competence is measured by your ability to communicate, empower your team, deliver the results, and manage resources and budgets. This is not by how well you code, even if you're a magnificent coder.

Successful PMs need to master these five competencies to be effective at their jobs.

1. Communication

There are a lot of people you need to be able to satisfy when you're the PM, and ensuring that you understand them and they understand you is important. You have your clients, project team members, project sponsor (more detail about them later in the book), and potentially other stakeholders as well.

All these groups are likely to be diverse, with different backgrounds and experiences. On the same project, you may have high school graduates and others with PhDs. The lines of communication must be clear with all of them, no matter what your own experiences and histories are.

With all the people involved in the project, you must be honest and direct without trampling over any relationships or damaging them. When you're able to communicate clearly, you can build more trust, which is a crucial building block to a successful project.

2. Negotiation

Along with all the different stakeholders, there are plenty of competing interests. Your client may want to add what they believe is a small upgrade to the project. Your team members may want to use a specific resource that will help them be more effective at their tasks, but the client's accountant doesn't want to spend the money. You may have competing departments within the client, or you might have battles from your own team between different priorities. Also, you have to manage all these competing interests within the constraints of the budget—picking two of fast, cheap, and good.

In other words, you'll need to be able to negotiate with and between all these interests. An effective PM is always searching for a win-win solution. Not everyone can get everything they want, but, hopefully, you can negotiate so they each get something.

Sometimes, you'll need to compromise. Other times you'll need to maintain a firm line about what can and can't be done on the project in a way that doesn't alienate the people you're saying "no" to.

3. Leadership

Commanding a team and being respected by all the stakeholders in the project are key to the project's success. Rather than using coercion on team members, strong leaders encourage and empower them to achieve the best results they can.

An effective PM can set their own ego aside when dealing with inter-personal relationships. How can they best serve the team and the project? They interact with the various stakeholders without breaking the bonds of trust, even while delivering challenging news.

Everyone can be a leader when they learn to nurture the right capabilities within themselves. You don't have to be an extrovert, just play to your strengths. More details about leadership in project management are coming in my other book. But, for now, it's enough to recognize that the more projects you lead and the more time you spend on learning these types of skills, the better off you'll be.

4. Organization and planning skills

As noted in the last chapter, these types of abilities have to be learned through experience and don't depend on the knowledge of others. Many project managers who are more familiar with traditional project management understand the importance of planning since the entire project must be structured out completely ahead of time before the work even starts.

Yes, these skills are crucial for agile success too. The PM must provide a framework at the beginning of how the project will be completed and plan the iterations as well.

5. Risk management

You learned many of the risk management details in the last chapter. Recall that the important component of risk management is being proactive. This is not waiting to respond until a molehill has become a mountain, but trying to stop an issue at the molehill stage before it can derail the entire project.

You must also recognize where and when you're not the expert. Consult people who've done similar projects when you're embarking on a type of project you haven't completed before. Check in with team members and review the literature if it's available. From time to time, you may discover that one of your stakeholders has some expertise you can tap into, but don't expect that to always be the case.

Vision

As the project manager, you're the one who sees the big picture and how all the smaller pieces fit into it. Stakeholders usually have a more limited view of what's in it for them, and your team members focus on the tactical tasks that they must perform to keep the project on schedule. Not only do you have the vision, but you also need to communicate it to all, then have everyone involved buy into it. All the stakeholders need to understand *how* the project will benefit them and the company, and how your vision will solve their problem.

Once everyone understands the direction and how it will lead to project success, the team can then begin to decide who does what and when. In a waterfall project, the PM leads the way on this task. In agile, with self-organizing teams, the PM may rather consult and encourage to ensure workers use their personal strengths whenever possible.

It's the project manager's job to empower their team to deliver results. That means allowing the team members to figure out the best way to tackle each task. Empowered teams have the maturity to be able to distribute the tasks evenly, the ability to make their own decisions, and the authority and tools to execute.

A team that can't manage its budget for a task and must ask the PM to sign off on the smallest items is not empowered. Such a team may feature pre-assigned roles that don't allow everyone to play to their strengths. They may also be unable to handle their own conflicts, so they must continuously request the project manager to get involved.

Ideally, the project manager is more like a coach once the team starts the project. They're available for consulting and can dive in to help solve a problem. They're managing the big-picture issues, such as potential risks and internal issues with the client, ensuring that all in-volved on the project are communicating and working together. The whole point of being able to empower the team is so they can handle the functional tasks by themselves once they understand the vision.

That doesn't mean they have free rein over the entire project, however. Teams need boundaries. For example, purchasing items under $100 may not require approval from the PM, but those over the threshold do. If the team wants to include an upgrade that will delay the timeline of the project, they should get the project manager's approval. However, daily decisions and relationship issues would be handled among themselves.

When a project manager doesn't give their team the power to make these decisions, it causes problems for everyone. The timeline will be slowed down because the members have to wait for the PM to approve every decision, no matter how minute. Rather than attempting to solve problems on their own, whether interpersonal or having to do with the task itself, the team members have no incentive to try to hash it out themselves. Instead, they'll depend on the PM to get them through the smallest ordeal. Because they're not trying to solve anything themselves, creativity will drop as well, since the only solution is the one the PM comes up with.

We've talked about how much more satisfied workers are at their jobs when they have some control over their day. A team that doesn't have the power to execute the project manager's vision doesn't have control over their tasks and schedules, so their productivity and job satisfaction drop as well. Being able to empower the team will make a huge difference to the success or failure of the project.

Leading with vision requires a dedication to continuous learning. Keeping up with technology and changing times is key to being able to execute that vision by having the appropriate tools ready for the team. If, at any point, you as a project manager can find something new that will help your team perform the work faster, more efficiently, and more easily, the better off everyone is.

It also requires some smart risk-taking on your part. Be bold when you're looking for breakthroughs. Playing it safe means fewer opportunities to excel. Taking intelligent chances is the only way to innovate. You'll need to game out the possible consequences of these actions first, which is the smart way to go about it. Maybe you'll need to consult with others who have been in similar situations, and perhaps you need to take your team's temperature first. However, don't ignore solutions or improvements just because they're not guaranteed to work. Think about them and the first-order consequences, then the second and third before you make a decision. Don't spend too much time in analysis paralysis, but go for it when the odds are with you.

Key Takeaways

Competency for project managers is measured in their ability to manage the entire project, not just accomplish certain functional tasks.

- You are the expert in project management, whereas your team members are subject matter experts on their tasks.

- Competent project managers are proficient in putting the pieces of the project puzzle together into a cohesive whole, and managing the relationships around it.

- Setting the vision, then empowering the team to deliver it is the most efficient way for a PM to work with their team.

In the next chapter, you will learn how to use the client's management to your advantage as a project manager.

CHAPTER

4

OUTSIDE THE PETRI DISH: MANAGEMENT CULTURE AND ORGANIZATIONAL BEHAVIOR

IT might seem obvious that the organization you're working with on a project would do anything in its power to make the project successful. As the project manager, you've set a clear vision for the project and how it benefits all the stakeholders, and you have benchmarks for success. Why wouldn't management be wholly captivated and willing to support you?

Unfortunately, you'll run into situations where it doesn't seem like management has your back. In fact, from time to time, you might find yourself under the wheels of the proverbial bus when they try to blame you for project failure! The more buy-in you get from management, especially the upper parts of the hierarchy, the more support you'll have throughout the project.

"Project management methodologies, regardless how good, are simply pieces of paper. What converts these pieces of paper into a world-class methodology is the culture of the organization and how quickly project management is accepted and used. Superior project management is attained when the organization has a culture based on effective trust, communication, cooperation, and teamwork." — Harold Kerzner, well-known management consultant and author

If only the PM's job were as simple as managing the Triple Constraint triangle! The project manager is the one who needs to build trust between the project team and the organization, so the team can receive the resources it needs for a successful project without too much interference from other stakeholders. When the team isn't supported, the PM will struggle to communicate, obtain the required resources, and encourage the team through the obstacles.

The most significant factor in project success is the organizational culture (Machado dos Santos et al., 2019).[6] The next most important is change management, which we'll discuss in a later chapter, and support from top management.

Organizational Culture

Also known as corporate culture, this is the beliefs, attitudes, behaviors, and values of the company's employees. Components of organizational culture include:

- Perspective on authority and leadership
- Shared vision and expectations
- Work ethic
- Social norms
- Incentive systems
- Code of conduct
- Risk management

For successful and mature organizations, project management culture is integral to the way they strategize and operate their business. It's embedded in the workplace, including the reward and motivation systems, and how the rank and file view leadership. A culture that recognizes and embraces project management is one that boosts the potential success rate of its projects. Its senior management comprehends the value of project management, aligns projects with business objectives, and demonstrates their commitment to their projects by investing time and money into them.

Senior managers in a strong project management culture demonstrate the type of leadership they want to see in all levels of the hierarchy. When they spot roadblocks in the way of a project, they come in and remove the barriers. Projects aren't an activity that's occasionally done outside the business plan or strategic objectives; they're an integral part of the daily work that occurs at the company, and their successes and failures are taken as seriously by the executives as any other business initiative.

A company's culture is its DNA, which shapes the behavior of its workers. Unfortunately, it's more fragile than human DNA because a mature project management culture can be destroyed with leadership that doesn't value or understand project management. Top executives forge the path for successful culture and can just as easily destroy it when they don't visibly support the projects and project teams at work.

As a project manager, when you're working for a solid project management culture, you'll know it because leadership has identified the responsibilities of your role and supports you in your work. Unfortunately, not all companies have a strong project management culture. Some may have weak cultures where project management isn't considered a key aspect of the business. However, with some prodding and building the case for project management, this type of organization can become one with strong project support.

More difficult are the firms where the company culture is strong, but project management is immature. These types of companies concentrate on functional tasks rather than cross-functional projects and don't understand the value that projects bring to the business. They may be harder to transform into a strong and mature organization because you'll be fighting the entrenched culture to make changes.

The overall project success rate in an organization reflects both its culture and the maturity of its project management function.

Values of the Project Management Culture

When the right values that support project management are embodied in the organization, there's a huge difference in how employees behave.

● Communication

When this value is present, workers have an open dialogue with each other and with their superiors. Management has an open-door policy, which will also lead to more cooperation. Communication and cooperation feed off each other in a mature firm. Discussions are held regularly, both within levels of the hierarchy and between them, and employees understand what management is thinking and vice versa. Status reports are quick and short because everyone knows what's going on.

When there isn't much communication, there aren't many people whom employees can talk to openly. Instead of short reports, there must be formal (and long) status meetings to catch everyone up to the latest data. All communication must be documented, which takes enormous amounts of time.

- **Trust**

  Employees who trust leadership and each other, and organizations that trust project management, are successful in their work. All levels of management are project sponsors (discussed in detail in the next chapter), and KPIs are aligned with the projects.

  When there is no trust, there's also very little success, competition among employees and management, less work completed, and more power grabs. KPIs are misaligned with projects.

- **Teamwork**

  Everyone feels free to share when there's a healthy attitude promoting teamwork in the business. Employees are committed to their own work, but also that of the team, and they leverage each other's strengths.

  Good luck getting projects completed when teamwork is not a core value! If anything gets shared, it's only for the employee's glory. There are no personal bonds between the workers or management; thus no one is committed to the team.

- **Cooperation**

  Here, you have support up and down the management hierarchy. Strategies include project management, and they are clearly communicated. Goals are also coherent and explicit.

  How does it look when there is no cooperation or support? "It's not my job" to ensure project success, so no one ends up doing the job. Not only are priorities not aligned with the projects, but there may be sabotage that tries to ensure the project does not succeed.

Conditions for Active Management Support

Given that top management's support is one of the critical factors in project success, you can see that it's necessary to get buy-in from the highest levels of leadership. Those lower on the ladder take their cues from their superiors. If the CEO can't be bothered and doesn't seem to think the project is important, so why should anyone else? Once the top of the pyramid is actively supporting the project, the lower levels will follow.

It's not enough for managers to say that they're only responsible for the project being behind schedule. They also need to demonstrate their interest. Ask the team what they can do to help, and, if there is anything, deliver on it. Employees are quick to notice when their management is all talk and no action. Project success depends on active support from all levels of the hierarchy.

Management must stand behind the team, whether they're progressing or not. No fair-weather support will make a project successful! Only 100% backing will allow the PM to complete it on time and budget.

Driving Active Support

The best way to ensure that project management is successful within an organization is to bake it into the company's business plan, vision, and mission. When integrated properly, projects align with the values and financials of the firm, where project success means company success. When senior executives have the incentive to support project managers proactively, there's little work on the PM's part to get their buy-in. The project manager, in this case, feels that progress is made easily and they can rely on the executives for the support they need.

Once an organization embraces project management into its cultural values, it can then set up a framework that helps them initiate the right projects and deliver them successfully. There are three levels of management: those at the senior executive level, senior management, and line management.

Senior executives working within such a culture demonstrate their commitment regularly to project management in a variety of ways.

● Align projects with business objectives

The more people have an incentive to support a project, the easier it will be to get buy-in! When it's clear how the project will support the business, everyone has a reason to support it.

When the project isn't aligned clearly with the business plan, it becomes easier for management to downplay the project as optional, or as something that's "nice to have." When it's necessary to achieve the goals of the firm, no manager or executive will want to downplay it or stand in the way of its success.

● Implement organizational and portfolio project management

Organizational project management is what the organization uses to align project (along with program and portfolio) management with its strategy and objectives, and customize practices to fit within the culture and structure of the firm.

A program consists of a group of related projects. Portfolio project management groups piece together all the projects and programs to create one single place for oversight and management of them, and adopt standardized governance across the organization.

Putting both these frameworks in place is a way for the organization to embed the principles of project management into its practices. It ensures that all projects across the organization operate under the same standards and are governed in the same way.

● Educated in project management and its benefits

When senior executives understand how beneficial project management can be for their KPIs, it becomes much easier for them to support projects actively. They are familiar with how the process works and when and how they can best provide feedback when appropriate. They're aware that obstacles and issues arise during the execution of the project and do their best to mitigate them so that the team can continue with their work uninterrupted. In addition, they understand the necessity for them to be visible to the project teams and others, so everyone in the organization knows they're providing support.

● Development and career growth

For project teams to work at their best, they must always be in the process of continuous learning. Providing training and development is a key method to demonstrate that the organization supports project management because they're investing money and time into it.

This type of human resource development also includes developing project managers and furthering their careers. Investing in PMs is a clear signal that the organization values them.

● Establish project management culture

Traditional company culture is sometimes at odds with project management, particularly when it's agile. Project teams value communication and feedback during the process—not just at the final delivery. Operating in short sprints and iterations may be uncomfortable for those who are more used to a traditional form of work.

Project management culture embraces these differences and encourages more feedback, not just within teams, but also with other departments and sections. Having a view of all the stakeholders involved in an issue is a project management staple that can also be extremely helpful in management, yet not everyone embraces it.

Senior management has its role to play in successful project management as well.

● Recognize positive ROI and achieve KPIs

If you don't recognize this alphabet soup, that's OK! ROI is an acronym for Return On Investment. Senior managers who seek higher ROI from projects are more willing to take on other projects and support the project teams.

KPI stands for Key Performance Indicator, which is management-speak for the metrics that senior management is often responsible for. Successful projects can improve performance, so smart senior managers will do their best to support a project's successful completion.

● Remove barriers to project success

The project manager can't do everything themselves. Sometimes, there are barriers within the organization that may otherwise prevent the project's success. Senior managers who are steeped in the project management culture and know how much successful completion can boost their metrics will remove roadblocks that could get in the way.

● Provide necessary resources

Project teams need more resources than just their brainpower. They may need materials, expert team members, or even just the opportunity to get their questions answered by an expert on the organization. Senior management is well placed to give them what they need.

● Establish guidelines and strategy

These are particularly important when multiple projects are going on at the same time. Projects should be executed in a reasonable timeframe that matches up with the company's business objectives. They may be unable to devote enough resources to too many projects at once, so they need to be prioritized.

The firm may also want to establish guidelines around what dollar (or time) amounts require management approval before changes can be made, what the PMs are and aren't responsible for, etc.

Senior management, with their overview of the entire company, can put these in place effectively, so all the projects can operate smoothly and similarly to each other.

Finally, **line management** is also key to the project's success.

● Provide skilled resources

The project team may need an expert from the organization to help them out or answer their questions, and line managers are the ones who can approve the worker shifting their time to the project. Managers who see that their superiors support the project and who have their incentives aligned with the project can then juggle the need for resources efficiently.

● Ensure commitments are met

Line management is responsible for ensuring that the experts are provided, along with other resources where appropriate. When they're committed to the project's success, PMs find they have all the resources possible.

If you, as a PM, encounter an organization that demonstrates most or all of the above criteria at every level of management, consider yourself lucky. You're working with a firm that values project management and will support you and your team in delivering a successful project. This is a mature company (in terms of projects anyway) that understands and already values project managers.

Gaining Management Support

Unfortunately, you might not be working with a mature firm. You may end up working with companies that don't provide this type of support, yet you still need the buy-in of management. How do you go about securing support from an organization that isn't already set up to do it?

If a project has been authorized and a budget allocated to it, then someone bought into the idea. Likely, it was someone in the upper levels of management because lower-level types typically don't have the authorization for it. However, once the executives signed off on the agreement, they may not educate the rest of management on why the project is important and why it should be supported. That part is now up to you. If you can find and develop a project sponsor (which we'll discuss more in the next chapter), that will be a great advantage.

On the other hand, what if you can't find a project sponsor and you're dealing with a less mature firm? You may need to do a little digging, and whoever authorized the project (or their assistant) will be the best place to start when figuring out what you need to know.

"Show me the incentive and I'll show you the outcome."
—Charlie Munger

1. Who are the stakeholders, and where are they in the company hierarchy?

Your team needs to understand who your project will affect and how the change affects them. Each stakeholder group likely has at least a manager or supervisor, and they may also have a person of influence who can help you reach buy-in with the group.

You also need to understand where they are in the hierarchy because, if the supervisor of one of the groups won't take a step without their manager's approval, you will need to know that so you can approach their manager first.

Suppose you've been hired by the Chief Communications Officer (CCO) of a company to develop a marketing dashboard for the communications analysts. Obviously, the communications analysts are a stakeholder group representing users of the dashboard. They may or may not report to the same manager because they could be in different functional areas or business units. You will need buy-in from their managers.

Who do the managers report to? With a flat structure for the organization, they could all report to the Senior Vice-President of marketing, who would report directly to the CCO. You'd also need buy-in from the senior VP; however, with more complex organizations, there may be an intermediate layer of management between the Senior VP and the CCO, both of whom need to be involved. Otherwise, the managers may not all report to the same VP, so you would need to add them in.

Are you finished with stakeholders? Maybe not. Who else uses the data from the marketing dashboard? Perhaps the communications analysts send out reports and the new dashboard changes to the creative team. You'll need to get buy-in from their manager too.

In other words, there are probably multiple levels of managers and stakeholder groups that you'll need to work with to increase your project's probability of success. Do your best to identify them and see where they fall on the org chart.

2. What are their metrics

This might be a little more difficult to dig up, but you can probably make some educated guesses if you can't get the hard data on how the various managers and workers are incentivized.

For example, the communications analysts may have targets for the number of pieces they complete during the week, and the manager's KPI is based on that. Senior VPs may have incentives tied to the budget, ROI, or a number of leads generated by a piece of marketing material, among other things.

If you know what their metrics are and what they get rewarded for, you'll know how to tailor your message to them. This way, you can increase the value for the stakeholders because it fits their KPIs and, as a result, they will be more than willing to support you.

3. Tune into station WII-FM

Take a step back and imagine that you're in the shoes of one of your stakeholders. They haven't been told why they should support this project, but have had it foisted upon them, at least in their minds. If you're leading an external team, they won't know you nor do they have any reason to trust you. Depending on the size of the company, they may have never met the person who signed off on their project.

They may not like what they're currently doing (that you're coming in to change), and they might have complained about it for some time, but it's familiar to them. There's a reason why people stay in their ruts, even when they're not particularly happy with them; change is unknown and, therefore, scary. These stakeholders know how to deal with whatever the current issues are, and now they've been told they have to take on new issues. If you were in that position, would you be happy about it?

If you've ever been in marketing, you may have recognized station WII-FM. It's an easy acronym to remember what you need to think about from the perspective of your stakeholder groups:

What's In It For Me?

Why should they accept this change that the senior executive has decreed for them? Because it makes their lives better in some way.

Making it faster, easier, and more efficient, it automates the boring, menial stuff so that they can get on with the more exciting parts of the job. It provides a higher ROI, doubles the lead generation from a single piece of marketing content ("collateral"), and gets rid of the bottleneck. Hence, the process takes much less time, and gathers all the data in one place so clients don't have to root through six different systems to analyze results.

For each stakeholder group, you need to explain *clearly* what's in it for them. It's not about the features of whatever you're putting in place, but how it improves their lives. For example, who wants a marketing dashboard? The CCO, but who else? Otherwise, people may wish to an easy-to-use system in which they can see how a single piece of collateral performed across all social media platforms.

They might want a system that tracks their ad dollars spent online with the number of prospects who clicked on the ad link. This is known as CPC or cost per click. Being able to track CPC across all platforms at once makes it easy to see where the company should be spending its ad dollars, rather than wasting people's time adding up numbers in a spreadsheet. Automating menial tasks so your analysts can do more powerful work is something people buy into.

This may not apply to the Senior VP, who likely doesn't care if the users are totting up numbers in a spreadsheet or using software. However, this one marketing dashboard means the six other pieces of software being used to deliver data to the spreadsheets are no longer necessary and can be reduced, saving huge amounts of time and expenses. There's an incentive for anyone with budget authority.

Once you can identify the people who need to demonstrate their support for the project victory, you can then make your case to them. Think in terms of **"head, heart, and hands"** (May, 2016).[7]

Communicate

It should be just like voting: early and often. It should also be taken seriously; however, make sure you and team members are regularly in touch while continuing to feed the reasons for what's in it for the stakeholders, which helps them understand why they should support the team.

It also sends the message that the project team values the stakeholders' input and concerns, which gives the latter a feeling of control and validation, making them more likely to support you.

Build the business case

Many managers, especially those with authority over their resources, need to see how the project supports their departments and goals. If they see how it will increase ROI, improve their KPIs, or help them retain more clients and increase client satisfaction, they will then have a good reason to support the project.

Do your best to build a case that supports all the stakeholder groups. Then, you just have one document that you can bring to meetings with all the different departments in play.

Document how the managers can help you in specific action items

You can sometimes win over the managers, but they may not be sure how they can demonstrate their support. Maybe you identified interdepartmental meetings where the manager can express their backing of the project and how they've devoted time or resources to it.

The more specific actions you can find, the better. People in management tend to be action-oriented; thus they feel better when they know what they can do to bring about the desired outcome—in your case, a successful project.

● **Measure progress with buy-in**

As the project's work commences, check in with your stakeholders to measure how much they're supporting and want to support the project. Once you've convinced them of how they'll benefit specifically from the work, they will believe that it's how they can achieve their goals. At that stage, you have their hearts. Once they join you in the work, even if just to provide you with the resources you need, you have their hands.

You also want to watch for stakeholders who *aren't* buying into the project. What are their remaining concerns? Was communication with them not persuasive?

If you're having difficulty with getting buy-in, even after you've done this work, ask yourself (and management) why. Do they think the project will shift their strategy or culture in ways they're concerned about? Are there financial issues on the horizon? Management may or may not be willing to share, but this is helpful information for you to know.

Organizational Change Management (OCM)

This is the approach companies use when they want or need to change their culture, such as moving from an inefficient or ambivalent project management organization to a mature one. A project manager leads many of these programs, but they require additional skills compared to a standard project that's not about changing the firm's identity.

As you've probably experienced in your own business and personal lives, human beings are very resistant to change. It turns out there's a neurological basis for it, making it potentially hardwired in (Gilbert, 2012).[8]

Essentially, the brain detects a change in the environment. That signal goes to the prefrontal cortex, which is where reasoning and higher functioning reside. This part of the brain compares that change with what's stored already in the mind and existing habits (basal ganglia). If those two don't match up, an error signal is then sent to the amygdala.

The amygdala is the area of the brain that was inherited from our reptilian ancestors and contains the fight-or-flight reflex, among other quick reactions and reflexes intended to get our ancestors out of danger. In other words, the amygdala can't tell whether it's fighting a sabertoothed cat or merely an information error; it only knows that the brain is under attack. So, it amps up the fear reaction, including stress hormones like cortisol, increases the breathing rate (so you can take in more oxygen as you're running away from predators), among other reactions.

This is the main reason why people resist change. Their brains think they're being chased across the country by a predator, which adds a degree of difficulty to the process! Unfortunately, it's not an appropriate reaction to change in the workplace, and trying to tell the amygdala that will yield poor results. However, that's what you're dealing with when you're leading and organizing OCM.

Many different frameworks for successful change management have been used in large and small businesses. Ultimately, OCM is taking the firm from its current state to its future state. When dealing with a culture that doesn't understand the value of project management right now, you're taking it from being ambivalent about projects or strongly immature to an organization that fully supports project management and is invested in the success of its projects at all levels of management.

1. Prep the organization for change

Here's where your vision as a project manager can be used to full effect. To convince the firm that they need to buy into project management, you must show them what kind of wonderful world awaits them on the other side.

What are the benefits of the strong culture that incorporates projects into their business plans and objectives? How does it make their lives easier or work more efficient and productive? Most importantly, you should think about how well-managed projects can ensure strategic goals are achieved.

Create that tension between what they have now, what is sub-optimal, and the future state. They need to understand that there is a gap before they can decide to change it.

You'll need to get buy-in from senior leadership at the beginning, which you'll then use to build at the lower levels of the hierarchy as well. Share the vision, so they can clearly recognize what they're missing by not building this culture. Then, develop the plan to bridge that gap. They'll see what's not working while also realizing they can have an improved future. With your help, they can get *there* from *here*.

You'll need to draw on your communications skills at this stage and throughout the project. Showing them that gap consistently and letting them know how they will cross it will help them internalize how change can and should be done.

2. Make adjustments to the operations

This is the most difficult stage because it requires you to fight that neurological resistance to change. It can be done, and putting yourself in the employees' shoes to comprehend what you're asking them to do is a crucial part of getting this step right. When you understand their fears, you'll know how to combat them effectively.

In this stage, managers must model the behaviors they want to see, which will help the firm bridge the gap. They'll need to clarify what's acceptable and what isn't. This needs to happen with your help because *you* are the expert in project management, and you can tell them what's effective in a strong culture.

Most importantly, the business structures and incentives must be aligned with project management success for change to benefit the company. Projects must become part of the business planning and be considered in the business objectives. Managers and supervisors also need to be rewarded for project success, so they can have more reasons to support the projects.

All while this is happening, you, as a PM, need to remind everyone of the vision regularly. This will include why they're being asked to overcome their natural resistance to change and do things differently. Change is hard, and all must have the vision consistently in their minds, so they can execute the tasks necessary to achieve it.

Workers should be empowered to decide how to make the required adjustments wherever possible. If they see roadblocks, they can devise a new method to overcome them. The more choice and control they have over their work, the less resistance to change that the management team will face.

3. Make sure it sticks

At this stage, the leaders need to "refreeze" the new habits and behaviors to ensure the new project management culture becomes entrenched firmly. They need to prevent backsliding when it (almost inevitably) occurs and make sure the changes are institutionalized in the culture, documentation, and procedures.

From time to time, they should review their progress to ensure the change is occurring on schedule and so they can spot any backsliding that may happen.

Key Takeaways

Having an organization that buys into project management at all levels of the management hierarchy is a major factor in the success of your projects. If that type of mature culture doesn't exist, you may need to lead organizational change or, at the very least, build management support for the project.

- Mature organizational culture prizes and supports project management.

- Immature cultures need change management to increase their buy-in for projects and become mature.

- The support of management is key at all levels—from senior executives down to line management.

- If you don't have this support, you can build it using different strategies with different management positions.

In the next chapter, you will learn how to use project sponsorship to help you win.

---

[6] https://**doi.org**/10.1590/0103-6513.20180108

[7] https://**www.dashe.com**/blog/the-importance-of-stakeholder-buy-in

[8] https://**www.batimes.com**/articles/the-change-management-life-cycle-involve-your-people-to-ensure-success.html

CHAPTER

5

PROJECT SPONSORS: THE KEYS TO THE PROJECT MANAGEMENT KINGDOM

HAVING a good project sponsor can make or break a project. As a project manager, it's up to you to cultivate these relationships to make your life (and that of the project team) so much easier. If you work for an immature organization, there may be no project sponsor, or the one who is present may cause more problems than they solve. There are ways to help sponsors become more effective and also some methods that can help you neutralize those that could be working against you.

Who Is a Project Sponsor?

This person "owns" the project, has the highest interest in it, and is responsible for providing resources and support to the project. Ideally, they would smooth the way for the project team to be successful, helping senior management buy into the project, and removing any barriers that stand in the way. They typically initiate the project and get involved from day one. Normally, they're the ones who select the PM.

They are the champion of the project within the organization. They achieve buy-in from management, ensure that the lower-level managers are on board with the project, and provide the necessary resources. There are three primary obligations for the project sponsor: governance, vision, and value (Malsam, 2019).[9]

1. Governance

- Initiates the project
- Defines roles and responsibilities
- Supports the project's organization
- Provides resources
- Escalates for issues the PM can't handle themselves

2. Vision

- Aligns business case with business objectives, strategy, and plan
- Defines the success of the project and how it fits into the business strategy

3. Value

- Manages risks
- Controls and reviews the process
- Assists with decision making
- Determines project quality
- Assesses progress
- Ensures the project delivers value with positive results

You can see there's some overlap between the project manager and project sponsor duties, but the sponsor is the one who is ultimately accountable for the success of the project.

Collaboration Between the Project Manager and Sponsor

Since the sponsor is the one who can make the PM's life either easy or difficult, the project manager must cooperate with the sponsor as best they can. The sponsor provides the environment in which the PM can either sink or swim, can muster the resources that the project manager needs, and takes care of internal issues that the PM cannot resolve.

It's also the sponsor's charge to ensure that the project team, including the manager, understands how their project will impact the business. They must know how the business case aligns with the objectives and strategies of the firm as a whole. Then, they will have a sense of pride in their work because it's a crucial component of the company's success. The project becomes meaningful in a way it otherwise may not have been.

Unlike project or risk management, there aren't any codified frameworks for steps the project manager should take to work with the sponsor. However, there are known activities that can increase the probability of a successful working relationship.

- **Meet early and often**

    Developing a good partnership right away will help both people in their quest for project success. The PM needs to update the sponsor regularly on how the project is progressing, along with any potential risks they see coming over the horizon. As the sponsor should be well-respected in their firm and be able to communicate well with other members of the management team, they can use this prior knowledge to prevent issues from overwhelming the project.

The more they meet, the more trust they will build in each other, which is a critical factor in success. Even though both may have busy schedules while the project is ongoing, continuing to touch base periodically must be part of the schedule.

Define rules of engagement

The sponsor and manager need to work together to understand how the project will fit into the business. This includes how they will work with each other and how the project team will cooperate with the sponsor and other employees of the firm. In this way, nothing will be omitted or left out of the planning because both clearly understand the structure of the engagement.

Discuss the roles and expectations

What is the PM responsible for? Likewise, what is the sponsor? These guidelines should also be laid out at the beginning. Nothing gets overlooked, and there are fewer problems later on.

Just as the PM takes on the tactical aspects of the project and allows team members to focus on daily tasks, so too does the effective sponsor avoid getting bogged down in the project details.

Once the guidelines for the engagement and trust in the team have been established, the sponsor does not necessarily need to be hands-on with the project. They understand that the PM will keep them updated on progress and escalate anything the PM can't handle, or that would be more appropriate coming from the sponsor.

There are several actions that the PM should expect from a **good sponsor:**

- Design a robust project charter with the project manager.
- Participate meaningfully during the kickoff meetings.
- Approve only frameworks and plans that seem feasible in terms of time, money, and resources.
- Evaluate project progress against benchmarks and guide the PM.
- Prevent scope creep.
- Celebrate reaching key project milestones with the team.
- Remove roadblocks where possible.
- Participate in after-project assessments and evaluations.
- Ensure successful hand-offs, so changes aren't dropped after completion.
- Sign off on the completed project.

The PM's ability and the complexity of the project should dictate the level of support from the sponsor. When the project manager is new to the type of project they're managing or the complexity is higher, the sponsor should be more involved. If it's not too complex, the PM may need more coaching. However, if it's higher, the sponsor needs to be more hands-on and involved directly, yet still maintaining the more strategic approach.

When the project isn't as intricate and the PM is experienced, the sponsor should be mostly hands-off and can rely on the project manager for updates. For more complicated projects, the sponsor should support the PM more and act as a consultant.

Most questions that the PM needs to ask the sponsor should appear early in the engagement, so the division of roles and priorities are clear. They need to ensure the sponsor will provide necessary resources, and that they understand their obligations.

Problem Sponsors, Problem Projects

In an organization with a strong project management culture, where everyone from line managers to top executives understands the benefits and why the project is necessary, you probably won't have many problems, if any, with your sponsor. The firm will have sponsors who are knowledgeable about project benefits, can articulate them and encourage buy-in among the other levels of management, understand how to work with PMs, and provide them with the environment they need to complete a successful project. These sponsors are vocal and visible with their support and will try to remove any impediments to project success. They want you, as a PM, to communicate with them regularly, avoid "shooting the messenger," and empower you and your team to get the work done.

Unfortunately, you will run into sponsors who... are pretty much the exact opposite of that. Typically, you'll see them in immature organizations. Sometimes they're just clueless, though often they will attempt to micromanage the project instead of empowering you to deliver the goods. If they can't empower you to do your job, either they picked the wrong PM, or the organization selected the wrong sponsor. In some cases, it's both.

They're like seagulls, and you want to avoid them whenever possible. According to Kenneth Blanchard, "Seagull managers fly in, make a lot of noise, dump on everyone, then fly out" (1985)[10]. They try to manage all the details of the project, despite how that's not their job. Since they're not the experts in project management, and you are, this could result in disaster.

If you have a seagull sponsor, you can look forward to a difficult life as a project manager because they'll engage in some or all of the following behaviors:

- Constant questioning or even ignoring your decisions as a PM.
- Micromanages the project.
- Doesn't understand the benefits of project management or the specific project you're managing.
- Discourages you from raising negative issues.
- Takes credit for successes, but blames you for failures.
- Changes priorities on a whim.
- Asks for long and detailed status reports frequently.

Build a Better Sponsor

If you see red flags waving when you first meet your project sponsor, and you're concerned you have a seagull, fortunately, there are some concrete steps you can take to try to develop them into a sponsor who will help you instead of hinder you.

Do a little research on them before you get started. LinkedIn is a great place to start looking because many professionals use this platform to network. Just be aware that if the sponsor has the Sales Navigator application, they can see the names of people who have viewed their profile.

See if you can find any commonalities: anything in your background that matches theirs. Even attending the same college, whether you did so at the same time, is a great starting point. Otherwise, you can see if they're engaged in any similar groups. This can help you build a connection with them, along with some trust, right away. This works with good project sponsors too.

Sit down with them to discuss the roles and responsibilities. You may find the use of a checklist helpful to ensure that you hit all the main points. Document the meeting and make sure the sponsor has a record of what you both agreed to.

You'll need to spend some time with the sponsor to help them understand what their role is and why it is important, along with why the project is necessary. If you had to develop a business case for senior management, you'd also find it helpful here. There are five methods you can use to influence them to become a better sponsor.

1. Communicate, communicate, communicate

To build trust, you may need to be in touch with them more often than you want or think you have time for. However, they need to see that you aren't trying to hide sensitive issues and that you're timely with potential issues as well—not delaying anything. This communication also includes regular update meetings.

2. Stay tuned to WII-FM

Here's where your research and business case really pay off. You should have a sense of your sponsor's incentives and interests, so you can appeal to what they care about. How does the successful completion of the project benefit your sponsor specifically? Not just the company as a whole, but *for them*? Continue to remind them of what's in it for them throughout the project—not only in the beginning.

3. Appeal to their desire to be a great sponsor

Most people enjoy mastering their position. They want to be good at it and have others view them as professionals. Let this innate desire help you work with them. Let them know how successful sponsors act and communicate. As always, tie it back to what's in it for them; how will being recognized as a good sponsor benefit them and their career?

4. Continue to focus on the common goal: a successfully completed project

Ultimately, the project is not about you as a project manager or them as a project sponsor. It's about the completion of a project that achieves its goals and objectives. As they say in 12-step programs: "Principles before personalities."

5. Avoid being condescending or using the word "should"

You're indeed the expert in project management, but you can still derail a relationship by acting as if you know everything and the other person knows nothing. Treat them as equals—remember that, likewise, there's a lot they know that you don't!

Do you enjoy it when other people tell you that you *should* do this, or *should not* do that? If you don't enjoy it, make sure you don't use such language with your sponsor either. Value and build a bridge with them; don't tear it down.

One method to get your point across without too much "should-ing" is to tell them stories of what you've seen work for you and others, and what hasn't worked in similar situations. Most people like stories and they can connect those dots pretty quickly. They won't feel like you're trying to decide for them or that they're giving up control, which everyone hates. In this case, you're not telling them what to do, but giving them options.

Keep them involved in the project as well. If you ask for their input on some higher-level decisions or to help brainstorm, they'll see that you respect them and their work. That helps them trust you while giving them more of a feeling of control.

Obviously, you would rather have a solid sponsor from the get-go; one who works in a mature organization and is developed enough to handle the responsibility of being the owner of the project. However, it doesn't always happen that way. It's up to you, as a project manager, to make sure you have the best sponsor you can, even if they didn't want or don't know how to do the job. Your maturity, as a PM, will be of great benefit when working with a sub-optimal sponsor because you'll need to take the lead in building the connection and developing the trust between you.

Key Takeaways

The project sponsor is the owner of the project and can champion a project to its success or completely derail it. As a project manager, when you run into a weak sponsor, you'll need to take over some of those responsibilities while you build them into a better one.

- The project sponsor is a critical factor in the project's success, and a good one will empower you and the team while removing roadblocks to success.

- As a PM, you should meet with the sponsor early to decide the rules of engagement and determine the structure of who does what.

- A good project sponsor does not micromanage the PM and adjusts their level of support according to the project manager's experience and the complexity of the project.

- Bad project sponsors may get bogged down in the details of the project and override the PM's decisions.

- If you end up with a weak sponsor, there are ways that you can work with them to build a better connection and develop them into the kind of sponsor that helps you rather than blocks you.

In the next chapter, you will learn how to close a project for success.

---

[9] https://www.projectmanager.com/blog/what-is-a-project-sponsor
[10] https://en.wikipedia.org/wiki/Seagull_management

CHAPTER 6

ALWAYS
BE
CLOSING

THE end of the project is just delivering the results that were defined in the scope, right? Not exactly. For the project to be completed successfully, there is a hand-off at the end. In addition, project managers need to deliver *throughout* the entire project process—not just at the end. There are many steps along the way that result in a thriving project, and too many wrong actions jeopardize that success just as an immature project management culture can.

Strong Execution

As a project manager, your job isn't over after you've planned everything out, at the beginning of the project, or after the sprint. Your job is to balance the supervision of the whole process and get involved where necessary while preventing yourself from being bogged down in the details or not empowering your team, so they must come to you for every small decision. Effective PMs do get their hands dirty, often with supervisory tasks like ensuring that someone follows up on the action items, adding needed resources, and the all-important skill of herding cats.

You have to stay on top of the project and make sure that deadlines are met. Since you may be the only one who can see how the puzzle fits together, you have to be in constant communication with the other pieces to make sure any issues are being handled appropriately or escalated to you when necessary. It's easy for team members to focus only on the task in front of them, so they need you to keep the big picture in view and help them align their work with the objectives of the process.

Just as the project sponsor is responsible for removing barriers to the project's completion and ensuring that management isn't interfering with the process, you need to do the same for your project team. They should be able to focus on the work while you handle external risks and barriers.

At the same time, you're working with people, so they must all feel comfortable coming to you with problems or issues that arise. That includes both stakeholders and members of your team. Having empathy and being able to listen actively are two key traits for an effective PM. If you ask your team where they're getting stuck and pay attention to the answer, it will be much easier to tackle roadblocks on their behalf.

Also, help your team look good in front of the stakeholders by supporting them in preparing presentations or other materials for meetings. It's not your job to create the designs or make the copies, but ensure there are no mistakes and everything hangs together coherently. Not everyone is skilled verbally, so they may need more assistance in putting the documents together. Your task is also to massage the tone of the message and level of details according to the needs of the stakeholders they will meet.

As a PM, you're also the head prioritizer for the project once the sponsor and stakeholders have agreed on the scope. You probably won't be able to deal with everything at once, so your job is to prioritize the top three or four tasks or goals that are most important to the project's success at any given time, and guide the team accordingly.

You'll likely have a lot of information coming your way, some of which is helpful, whereas some not as much. In other words, you'll be taking on the role of a filter as well. If the info assists the team, you can let it through; otherwise, scrap or archive it if you think you may need it later. This allows the team to focus on the work without too many outside distractions because you will take care of it.

Stress is something that many PMs experience, particularly with all the distractions they're faced with. To prepare for stress, make sure that you have time to exercise, provide yourself with a pleasant night-time environment so you can sleep well, and eat nourishing food that doesn't compromise your health or immune system.

One of the most effective ways to reduce stress is to delegate. When you've surrounded yourself with honest, trustworthy people, you won't have to do everything yourself. Find ways to coach people, so they will take on more duties without being overloaded, thus taking up some of your burdens.

As the ultimate decision-maker for the project team, you may sometimes be the one making the decisions, usually under conditions in which you don't know all the information perfectly. However, it will help both you and the team if you coach them through decisions when you can. It also helps team members hone their critical thinking skills. Note that it may mean a little more time spent on your side because you may have been able to make that choice pretty easily yourself. In the long run, however, having team members who can make those kinds of decisions will benefit you and free up your time.

Take comfort in how, if you can manage all this, everyone will hate you equally at the end of the project if you've done your job correctly!

"The best is the enemy of the good." –Voltaire

Given that we are all humans and, therefore, perfection isn't feasible, a project manager needs to focus on what's good enough. If you're continually trying to achieve perfection, you'll never be able to release something that satisfies the requirements. Besides, you will have a lot of items on your plate, and trying to give 100% on all of them won't be possible. Work on making it good enough, then move on. Remember that you are not delivering results to yourself, but to the stakeholders and their expectations.

You're probably familiar with the Pareto Principle, which says that 80% of the result is driven by 20% of the contribution. It's no less true for project management, which we can illustrate with the following three aspects.

- **Cost**

  Most of the expenses on any given project are due to salary or contract wages. People are the most pricy contributions to the budget, so make sure you're getting the bang for your buck. Choose talented, mature team members who will provide value for their labor.

  This also means that it's not the resource issue that's your enemy on a successful project. It's *time*. The better you can manage time-lines, the more likely you will be to achieve your objectives. When you get your tasks done when they should be, you set a good example for the rest of the team.

- **Contributors**

    When it comes to your team, 20% of members will produce 80% of the work. Focus on these high performers and provide them the resources they need to perform at their best. You probably won't have enough time and money for the whole team, so make sure you are prioritizing your *star* workers.

- **Results**

    If you manage to get 80% of the goals targeted in the scope of work, congratulations! You've *completed* your project successfully. The key here is to focus on the activities and goals that have the highest return on investment. These will be ones that provide the biggest impact and get you closest to the overall goal.

With all of this, the joke is that projects are always 95% complete and never finished. Therefore, strive to be that rare PM who finishes one project and moves on to another. Hand it off, even if it's in project phase one, so you can then move on to project phase two.

Closing Out

Simply handing over the deliverables isn't always enough for a successful project. You want the changes you've made to stick, and for them to do so, you'll need to focus on closing tasks. You, the team, and the stakeholders have spent a lot of time and energy on the project; what a shame to waste it by not genuinely following through at the end to ensure the project results are safely in the hands of those who will benefit from and use them.

Not only that, but the team's credibility and reputation could be damaged if something goes wrong after the delivery. If the process change you implemented, for example, doesn't stick, it could be viewed as the team's fault. There are a couple of things that can go wrong when the closing process isn't handled correctly.

● Orphan item

If the receiving team or department doesn't undergo the necessary training and awareness for the project's deliverable, they won't be able to use it. If you (or the project sponsor) didn't work through the buy-in process with the users of the deliverable, they wouldn't use it because they don't see the need for it, and probably feel like it's been foisted upon them.

By contrast, when the users know why they're receiving the deliverable, have been trained on how to use it, and had input on the process, they will be more likely to take charge of it after the hand-off is done properly.

Imagine you bought a brand-new computer, but no one at the store or the manufacturer could fix it when something went wrong. The project team may have developed it, but there was no training beyond the team on the new item. In this example, the store and manufacturer dropped all responsibility for it beyond adding it to the shelves.

● The never-ending project

In this case, the organization continues to hold the project team accountable for operating and maintaining the deliverable, rather than the departments who should be taking responsibility. The receiving departments don't train or task their members to understand and operate the deliverable, leaving the project team responsible without the necessary capacity and skills to maintain it.

Suppose you bought the brand-new computer, but every time you have a question about it, the manufacturer would send you back to the project team. They built it, but they didn't load on the software or had anything to do with its creation, so they cannot answer your questions.

Also, as we've discussed, scope creep and the additions of extras keep the project going long after it should have been completed.

To avoid these scenarios, the project manager needs to close out the project thoroughly and ensure the transfer from the project team to the department is handled properly. There are three steps, which may seem obvious or trite (Aziz, 2015).[11] However, if you don't carry out and document them correctly, you could end up with unearned blame or reputation damage.

1. Declare all work has been completed

The PM should check in with the project sponsor and the customer and make sure they approved the delivered work. Any contracts (such as procurement) should be reviewed for completion and ensured both parties had executed them to the full extent of the agreement.

2. Assurance that all project processes have been achieved

Review the process and governance documents to ensure that they've been executed. Validate the achievement of business case objectives.

3. Formal agreement by all parties that the project is complete

Recognize that the project is complete and the transition of the deliverable to operations. Free the resources up from the project to begin work on other projects, or return to their functional teams.

Capture the lessons learned. This is especially important for the team members, who may have been unable to have a perspective on how the whole project operated. Recording lessons learned helps both the team members and stakeholders implement improvements in the future and avoid losing valuable information.

Sometimes You Have to "Take the L"

Just as you don't want to accept unwarranted blame, avoid doing the same to others. As you know, the buck stops with you as the project manager. If something doesn't go right and it's under the team's umbrella, you may have to *"take the L"* (loss).

Team members will get things wrong from time to time, as you do too. Good leaders look for the lesson in the mistakes and avoid overly harsh punishment. Empathy and rewards for good work go a lot farther than coercion or threats.

Will you need to call people out from time to time? Probably. If there's a lesson to be learned and the whole team needs to hear it, you often do need to put that person on the spot. However, you don't need to shame them in front of the team either.

When something goes wrong and you can fix it, do so. Even better is if you can coach the person who got it wrong on how to fix it. Either way, avoid the blame game and focus on getting the process back on track.

People are your biggest expense, and they're also your most significant resource. Deploy them to the best advantage of them and the team, with coaching to help them over the hurdles. When the team does hit a roadblock that the sponsor couldn't or didn't remove, don't panic. Work with your players to get around or over it if you can't move it.

Being a project manager requires a degree of maturity, which means you must avoid scapegoating. It doesn't matter who—don't blame the client, sponsor, team member, or any stakeholders. Even if you think only your team will hear you, trust me, word gets around. Most often, it doesn't matter how you got to where you are now. What matters is the situation you're in and how you plan to get out of it and put the project back on track, so you can achieve your goals.

Key Takeaways

Completing the project on time is what you need to focus along the way, including how you handle the transition from the project team to operations.

- There is no such thing as project perfection, so work on being good enough and leverage the 80% to drive results.

- Handing off the deliverable properly affects the reputation of the team while ensuring that the work and lessons learned don't go to waste.

- Sometimes, you will need to take the loss, so learn from mistakes and move on without spending time shifting blame around.

In the next chapter, you will learn which technical skills you need to master.

---

[11] https://www.pmi.org/learning/library/importance-of-closing-process-group-9949

CHAPTER 7

TAKE COMMAND
OF THE TECHNICAL
SKILLS

WHILE it's true that project managers don't necessarily need to have the type of in-depth skill sets that members of the team do, you still need to master the capabilities required for an effective PM. The management of a project has its own requirements, which is where you are the expert and why firms want to hire you. Your team members are the masters of their specific domains, and you're the one who combines them and creates a structure for them to work at their best. However, your job doesn't end with organization and planning because you're the one who's held accountable for the success or failure.

Having said that, the more you understand what your team members are doing, the better. That will help you give an appropriate amount of time to the tasks and hire the right people for the work. It will also be easier to develop the project plan and evaluate performance when you have some knowledge of the benchmarks. If you work in software development, for example, knowing the fundamentals of coding will help you manage the project.

When there's a technical challenge, you'll be better equipped to step in if necessary. Ideally, you'd have someone trained to do that or learn on the job when needed. However, everyone, including the project manager, needs to do what they can to achieve the goal. If you were unable to hire as many team members as you would have preferred, you might be the one to patch over the gaps in skills where the others cannot.

Technical Skills

You probably are already aware that project management tools and approaches change regularly, so you need to stay on top of the project management industry and ensure that you're using the right tools for the job.

• Project management methodology

The history of project management, at least formally, begins with traditional, or waterfall, management. In this style, the PM plans the entire project out from beginning to end and manages with a top-down approach. One project may take months or even years to result in the deliverable. The stakeholders have a chance to put in their two cents at the beginning, and only receive status updates until the end. Most tasks occur sequentially, and the next task must wait for the previous one to finish before it can get started.

The software industry developed a different methodology known as agile, though many other industries have discovered its benefits. It can respond much faster to changes in the environment and provides deliverables much quicker. The work occurs in iterations—or *sprints*—of no more than a month long, and the stakeholders stay involved during the entire process. The PM is often more of a player-coach than a top-down manager.

Some projects, especially those that can't tolerate much change in a plan, are better off with traditional, such as the construction of a bridge. Others that require fast adaptation, like IT projects, are better off with agile. As a PM, you'll need to understand the strengths and weaknesses of each and when they're used best.

The vast majority (97%) of software development firms use agile frameworks, though not all of their project teams are truly agile (Barker, 2019).[12] In the next section, we'll discuss the more popular frameworks that exist within agile.

● Measurement approaches

Earned value management is a tool designed to measure the performance of the project. It enables the PM to see where the project has been, where it is now, and where it's going. Earned Value is also known as the budgeted cost of work performed (BCWP); planned value is where the project should be in terms of the budget as of today. By comparing the planned value, earned value, and actual value of the work up to that point in time, the project manager can estimate completion times and burn rates. It doesn't work on every project, and the data must be accurate for it to work correctly.

Other approaches to the question of "Where are you on the project?" include using incremental milestones or tracking when the project reaches a certain point compared to the expected achievement date. If there are no milestones then a "start-finish" measurement allows for progress only when the project is complete.

Others more qualitative (and therefore less attractive to PMs and the stakeholders) include measuring the level of effort and individual judgment. On some projects, particularly the more complex ones, it may be difficult to use any of the other methods. In some cases, a combination of approaches will work best.

Whenever possible, try to quantify the progress as much as you can, even when the measure is not precise. This will help you protect yourself because a measure is a fact that cannot be doubted. Opinion, however, can change quickly, especially in adverse situations when people try to blame others.

Tools

The more tasks can be automated, the better. Rather than wasting valuable time manually updating spreadsheets or Gantt charts, the team members and project manager can free up their time for decision making and problem solving.

There are many project management software tools available online that allow the entire team to collaborate, either online or in-person, from anywhere in the world. Any changes that someone makes to a task flows through all related items.

You may have heard some of the more popular ones, such as Trello, Asana, ClickUp, and JIRA. Most of them provide the ability to automate "old-school" tools such as Gantt charts, Work Breakdown Structure (WBS) diagrams, timelines, and mind maps.

Agile Project Management

Rather than being task-based like traditional project management, agile is focused on principles. In fact, there's a 12-statement manifesto that guides the thinking behind all agile projects. Four core values underpin every type of agile project:

1. People before processes.

2. Working prototypes instead of extensive documentation.

3. Collaboration with the client.

4. Adaptation rather than sticking to the plan.

These different values mean additional skills for project managers who are more used to waterfall management. Being able to communicate is much more important because the agile project manager is always in touch with team members and stakeholders. They need to know how to talk to people of various educations and backgrounds.

An agile PM is much more of a coach than a manager when dealing with their team members. Their job is not to plan everything out and determine who does what and when. Instead, they're there to empower their team members to take on the challenges and organize themselves.

The core values, especially of communication and collaboration, are fundamental for project managers to embed in their work. Even waterfall managers will see benefits when they coach their team members instead of continually managing them down. As a PM, you hire people because of their talent and ability to do the work, so don't interfere with that by micromanaging or commanding instead of encouraging.

Some types of agile don't even have a role called "project manager," though, in practice, the PM is still working with the team.

The most popular agile frameworks you may have heard of include the following.

- **Scrum**

  Over half of software development companies use this style, named after a rugby scrum (Barker, 2019).[13] The Scrum Master acts as a PM to a large extent and is a player-coach within the team.

- **Kanban**

  In this framework, the project manager can organize and track the progress using a visual Kanban board. It's flexible, and the guiding principle is to manage the flow of work. This will, in turn, allow for changes along the way, resulting in high-quality deliverables.

- **Six Sigma**

  Usually used for process improvements, Six Sigma reduces the probability of defects and minimizes variability. It can be applied for a wide variety of processes—not just those in manufacturing, though that is where Six Sigma started.

  Like Scrum, there are no project managers in Six Sigma. Leaders ascend from yellow belt as complete novices in the framework up through white, green, and black or master black belt. There are specific project items and experience required to move from level to level.

- **Extreme programming (XP)**

    This type of agile is explicitly aimed at software and providing high-quality applications that have been tested robustly. It also makes life better for the development team. This framework uses test-first programming rather than the usual process of writing code and then testing. It also involves pair programming, in which two people sit at one computer and code.

    There are no specific roles on the team because everyone is specifically cross-trained. However, new teams may bring in a coach who's experienced in XP to develop team members, which may be an appropriate role for a project manager.

You will find more details about agile approaches, principles, and success factors in my book *Become an Agile Project Manager*.

Embrace Automation

We touched on automation in the software tools discussion. Before all the software was widely available, many PMs hand-coded their Gantt charts and entered data manually in other tools, such as spreadsheets. Project managers who started earlier may be used to all this manual work, but there are distinct benefits to automating as much as you possibly can.

- **Highest and best use of resources**

    Going back to the Triple Constraint from the beginning of this book, every PM is limited in their resources, especially in terms of time and money. We also noted that people are both your best and most expensive resource. Using these precious resources to enter information manually or do other tasks that a computer can do better is simply a waste.

You weren't hired to update data; you were hired to oversee the entire project and, more importantly, make decisions and solve problems. You also didn't hire your team to work through spreadsheets. You hired them for their expertise, knowledge, and ability to get the work done.

Use your resources wisely, making the highest and best use of their time and your time.

Omitting omissions *(and other errors)*

Even the most detailed manual entry clerk can transpose numbers, leave out a part of the formula that causes everything else to be wrong, or forget a crucial step in their work.

A computer does not make such errors. The applications need to be programmed correctly, of course, but once it's set up for success, it won't fail, get tired, or take a break right when you need something to be updated urgently.

Smoothing workflow

Many detailed procedures and checklists can still fail or cause problems if the hand-off from one task to another is delayed or omitted entirely. Sometimes, workers may not understand who they're supposed to tag for the next step. In other cases, they may not be in the office, or some other common circumstance occurred.

Project management software allows for tasks to be owned, so when one is completed, the next owner can be notified automatically that their step is ready. Recent applications have advanced communication and storage tools that bring cooperation to another level of efficiency.

Backup Your Backups

During the planning phase, you should have identified some potential risks to the successful completion of the project. However, there will always be some unknown threats that no one sees coming, and you will need to have a flexible contingency plan for them. When they arise, stay calm and feel free to use your team and stakeholders to solve the problem.

Power outages and server issues are problems that everyone deals with from time to time, so your information needs to be backed up periodically. Most software providers offer this service, but you can also consider backing up on an external hard drive or some other source. What will you do if there's a natural disaster in the area? What if a team member is out sick for a significant period?

These are the types of things that hopefully won't happen, and you may not activate a plan when you don't need it. However, you will still need to have some ideas for what you'll do if they occur. For example, you may have other potential team members who can fill in for someone who will be out for a while. You wouldn't hire them just in case, but you would have their contact info in your proverbial back pocket should the need arise.

Always assume that something will go wrong because they will! Hopefully, whatever happens is not catastrophic to the project and, if you're lucky, you'll just have some mild irritations to deal with. However, you can't always count on such situations that happen only to be mild when wanting to ensure your project is a success. Thinking ahead and looking through the consequences of those consequences will help you manage through the ordeal.

"It must be remembered that project management is first and foremost a philosophy of management, not an elaborate set of tools and techniques. It will only be as effective as the people who use it." — Bryce's Law

Know Your Business

While having a fundamental knowledge of the project tasks and being an expert in project management is necessary, it won't be sufficient. The PM also needs to understand the industry in which they're working and the specific client who hired them for that project.

The PM must have a good grasp on the business case and how it impacts the organization and its business objectives. That helps the project manager assign priorities and know which of the two they're going to deliver—that being fast, cheap, or good. They understand the kinds of results their clients want to see, and can explain such to their team members as well.

To do that, however, the PM needs to understand the business itself. Who are its customers and what problems does the firm solve for them? What are the competitive advantages that give them a toehold over their competition, and what are the potential threats to the business? Even if your insight can't get too deep into their financials, you still need to understand where much of their profit comes from, along with their marketing, sales, and distribution efforts. This type of knowledge will also help you uncover the likely risks with the project, so you can plan to mitigate or avoid them if possible.

Knowledge of the industry is also helpful because you have an overview of the key players and their advantages and disadvantages. Step into the client's shoes and see who they would like to hire and the kinds of backgrounds they might prefer in a project manager.

Suppose you're looking for a project in the automotive industry in product development. If you were hiring a PM for your firm, would you rather employ someone who has worked in automotive before, or one whose specialty is in financials or IT?

You may be wondering how you can absorb this knowledge. Much of it is acquired over time. If you're a project manager who works within a specific organization, for example, you'll understand the industry better as you work on more projects.

Whether or not you are already working for the organization, you can always ask your project sponsor to help you fill in the blanks. Ask why this project specifically, about its timing, and how they expect it to impact the bottom line and their goals.

You'll also need to look for opportunities to learn about a particular company or field, depending on whether you plan to work for different organizations or just the one. If you don't have a chance at your current firm, consider volunteering in your community to get exposure, more knowledge, and more experience.

Key Takeaways

The project manager role requires a fundamental understanding of every position in the project, even if they're not subject-matter experts on the details. However, the PM must also be a master in project management and be able to choose the appropriate tools and techniques for their projects.

- As a project manager, you'll need to know the differences between waterfall and agile, which methods should be used on which projects, and stay up to date on performance measurement and other tools.

- Agile is a flexible methodology that allows for rapid changes and has been used increasingly by organizations all over the world.

- Automation helps PMs make the highest and best use of all their resources.

- Backing up information and creating contingency plans is necessary for every effective project manager.

- To be effective, PMs must also have a solid knowledge of the company and industry they're working in, so they can make the right decisions about the project.

In the next chapter, you will learn about taking your project management abilities to the next level.

---

12 https://**betanews.com**/2019/05/07/state-of-agile-report/
13 https://**betanews.com**/2019/05/07/state-of-agile-report/

CHAPTER

8

LEVEL UP
YOUR
PM GAME

RIGHT now, you should have a pretty good understanding of *what* effective PMs do. You've probably identified some areas in which you need to improve your knowledge or experience. In this chapter, we'll be talking about *how* project managers work to be more potent in managing successful projects.

Act with Integrity

Project managers can be seagulls, just like project sponsors or line managers; they could fly in, dump all over everything, then leave. Those types of PMs will find it difficult to maintain a career because word gets around. The smart and productive people who make up the team don't want to work for people like this, and they will turn down jobs. Project sponsors who hire PMs want a track record of accomplishments and good work habits.

The project manager that everyone wants to hire or work for is one whose career is based on honesty. This is the foundation for trust and relationships. People want to work with those who have strong morals and ethics and demonstrate them every day in their work. Empathy and interpersonal relationships are more important for leaders today than ever before, and only honest people can be authentic and genuine in their communications with others.

As a project manager, you're the role model for how the team behaves. If you don't act with integrity, neither will they. If you don't show that you're ethical in everything you do, neither will they. For the project to flow according to schedule, the team must work together. If they don't, there's very little chance of project success. Rather than completing their work, they'll end up shirking responsibilities. They lose the drive and motivation to make the highest quality deliverable possible because, with an unethical project manager, they will have no incentive to do their jobs. Team members with high morale often stand against unethical managers, and a lot of energy goes into internal fights for what is right and why it is important.

Doing what you said you would do is crucial for building good connections with your stakeholders and team members. If you say one thing and do another, why should anyone listen to what you have to say? Why should they trust you? Practicing what you preach is a key element in developing a strong and mature team. That's the kind of reputation that project managers want to have. When word gets around that you're a loyal PM who acts with integrity, you'll have plenty of offers from sponsors and colleagues who want to be a part of your next project.

Often, when you're working on a project and feeling stressed, you'll be tempted to bypass principles or rules and morals set by the company. However, a PM must act according to ethics and rules. If you try to manipulate results, hide the bad news, or make excuses because "everybody else does it this way," no one will want to work with or hire you. It's selfish and irresponsible to set aside ethics. People see that you act in ways that would only benefit yourself, not team members, or the project as a whole when problems arise. And everyone knows that complications are common.

Reduce Red Tape

You'll likely find more bureaucracy in immature organizations. It's the project sponsors and senior management who don't really understand how projects work and why they're essential. They demand extensive documentation, project plans, status reports, long meetings, and a commission to decide everything, thus there ends up being no personal responsibility. Though, in all honesty, PMs can be bureaucrats too and allow the projects to be wound about with unnecessary rules and regulations.

When you can remove the bureaucracy as far away from your team as possible, you free them up to do their work. Automate as much as you can to create project documentation. Remember that you're fighting against time and have limited resources, so make the highest and best use of your resources that you can.

Be Organized

OK, you don't have to apply Marie Kondo's style to your workspace if that doesn't work for you. However, the more organized you can be in your work, the better. Don't let mess (either physical or mental) interfere with the smooth operation of your project.

You're the standard-bearer for your team members, just as you are when acting with integrity. If you're not organized, they have no incentive to be either. More importantly, if there's anyone on the team who needs to know where everything in the process is and where it should be, it's you. How much progress has been made since the last milestone, and is there a variance between that and the plan?

Tracking progress and performance is critical for success, so you can make small tweaks along the way, rather than a huge course correction if the variances are allowed to grow too large.

Time management is another component of being truly organized. For the team to do their best work, they can't be stressed or under too much pressure. Your job is to balance this with the needs of the project for productive hours as best you can. You've hired great people, so you shouldn't have to push them very hard. Guide and help them set priorities for what must be done on any given day, then let them figure out how to achieve that. It does require you, as the PM, to set the priorities effectively and communicate them.

Understanding People and Psychology

To get the most from your team players, it's helpful to know a little bit about what motivates workers in general and how each person works best. Can you be an effective PM without understanding psychology? Yes, if you can hire good people who are driven to do a good job. But leveling up your game requires more knowledge about what makes people tick, so you can leverage that to get the best work out of your people without stressing them out. Fundamentally, great people want to do a great job, so if you can help them do that, you will end up with excellent results. But there is more to it than that.

From reading this book, you know a few things you may not have before. We discussed that fear of change is built into the human psyche. Messages travel between two parts of the brain to compare what the brain expects from previous experience to what it sees now.

If those two things are different, the mind then believes that it's under threat and activates alarm systems. You know that you'll need to focus on everyone's favorite radio station, WII-FM, to help them through any changes by demonstrating how it benefits them.

People also have different drivers of behavior. You'll find some people are motivated solely by money and get the most out of monetary rewards. Others need to be praised regularly for the good jobs they do, so they can feel appreciated. Some people just want the facts and desire to find out the truth of the matter. Understanding how your team members are motivated will help you provide rewards for outstanding performance. Not everyone gets a trophy in project management, but you do want to recognize those who go above and beyond.

If you want to understand people's personalities, there are different ways to compare and contrast, some of which you're probably familiar with, or have at least heard of. Before assessing your team members and stakeholders, you have to understand yourself, who you are, what motivates you, what you dislike, and how you can improve your character.

The following measures are relative. Nobody is 100% one personality type, and nobody acts the same in all situations. However, certain patterns of our behavior may be dominant. It often happens that one is neutral; for example, someone could be neither an extrovert nor an introvert. The good news is that no personality is better or worse than another because each has its pros and cons.

Myers-Briggs (MBTI)[14]

This test is very popular. It measures personalities along four axes, meaning there are 16 possible personality types within its chart.

1. Extraversion/Introversion

They measure whether people are more comfortable in a crowd or by themselves. It's common to believe that introverts are shy, but they may be quite outgoing when they're with people they know and trust. They are often great analysts. Extroverts, on the other hand, find it easy to speak to people, though they may often overlook important details.

2. Sensing/Intuition

How do people take in information? The ones who are higher in sensing rely on facts and what they can see and touch. These are the team members who enjoy a hands-on experience. People who score higher in intuition; however, focus on patterns and expressions; they enjoy learning and talking about abstract theories.

3. Thinking/Feeling

This axis is based on how people prefer to make decisions. Interestingly, neuroscience has discovered that our subconscious is the one making the decisions, and we justify them after the fact.

Those who score higher on thinking weigh out the advantages and disadvantages of the options before making the final decision. You may see them listing pros and cons, for example. The members higher in the feeling attribute put more emphasis on whether the decision will promote harmony in the group and how others may feel about certain decisions before making theirs.

4. Judging/Perceiving

Don't get tricked by the words on this scale because they're misleading. A person high on the judging scale isn't necessarily judgmental; they merely prefer some structure and closure with their activities. By contrast, high perceivers are more flexible and open and don't necessarily focus on finishing a task.

The MBTI may be difficult for you to use in practice due to its high number of possibilities. On the other hand, many people are familiar with this particular personality test and may already know which type they are.

DISC[15]

There are only four personality types in this assessment. You can think of it as a 2x2 matrix, with one axis being activity (fast-paced vs. cautious), whereas the other is a continuum of trusting vs. skeptical.

1. Dominance

Fast-paced and skeptical, the D team member doesn't want to hear a long story. Let them know what you want from them directly. They may also be innovative and often argumentative.

2. Influence

These team members are fast-paced and trusting, also making them persuasive and emotional. They're the peacemaker of the team. While they may sometimes be more concerned with popularity than the right decisions, they encourage the rest of the team.

3. Steadiness

Trusting and cautious, these team members are strong team players and good listeners. They tend to struggle with change, but they are still very dependable in their work.

4. Compliant

Don't be fooled by the name, as these team members are merely cautious and skeptical, not necessarily submissive to authority.

They're logical and analytical. Though they can get easily bogged down in the details, they don't like to argue either.

The DISC system has the benefit of being easy to remember, and you can probably eyeball your team members and determine where they would fall along each axis without too much effort.

Big Five (OCEAN)[16]

Psychologists often consider these the five factors—dubbed by its acronym "OCEAN"—determine a person's behavior most above other tests. As with the other tests described above, it is believed that people exist on a continuum for each of these factors. Understanding where they are on each aspect will help you determine how to work with them more effectively.

1. Openness to experience

This includes creativity and intellectual curiosity. Creative people are often good problem solvers because they can think outside the proverbial box.

2. Conscientiousness

This aspect measures how likely someone is to finish what they start, how responsible they feel toward their work, and how productive they may be. You definitely want people who are high on this scale on your team.

3. Extroversion

This scale is defined a little differently from the MBTI scale, though its opposite is still introversion. This scale measures assertiveness and sociability.

4. Agreeableness

How much does someone trust other people? Are they compassionate and respectful? Having a team in which the members are higher in agreeableness will encourage them to work together.

5. Neuroticism

This trait looks at how much the person's tendencies may lean them toward developing depression or anxiety.

In reality, you won't have time to delve too deep into your team players psychologically. However, you can get a reasonable understanding of whether they play well with others, run with scissors, and how they can be motivated during the time you spend with them.

If you recognize specific patterns in their behavior, you can reallocate tasks. One team member may be fit for gathering and processing data quickly. In contrast, another may be better for analytical work and decision-making, and someone else may be best for persuading others. In other cases, someone could be a great worker who needs just a few instructions, whereas another person could be a great negotiator who can improve harmony among the team.

Set, Manage, and Never Give Up on Your Expectations

People work best when they know what you want from them. At the outset of the project, this may be more difficult because it will evolve as you settle in and get to know all the players better. You also need to be firm on the expectations you have of the stakeholders, particularly on agile projects where you will be seeking feedback regularly. Also, set the expectations as early as possible in terms of what they can expect from you and the team.

When people come to you with questions, make sure they're within the scope of the work you agreed to. Otherwise, you could end up with the dreaded scope creep. If it's not your area of expertise and you don't know the answer, bring in your subject matter masters and let them handle it. If you don't have one at hand right when someone's asking you the question, that's OK. Just make sure you understand the question, follow up with the expert, then get back to the person with the question in a timely fashion.

You know you can't deliver the moon, so make sure you don't promise it upfront. That makes managing the expectations much more comfortable. Your communication skills are helpful here because you may need to adjust the deliverable or action as you go along. Although you may do your best to mitigate all issues, not all of them will be solvable, and you will need to communicate that. Again, do so in a timely fashion, not waiting until the end of the project and delivering something different from what was promised.

Don't let go of your expectations. If you set the standard for quality of work and it starts slipping, don't let it go. If the team has a required check-in with a stakeholder regularly, ensure it doesn't fall through the cracks. That includes any standards you've set for yourself—don't allow yourself to fall below them in the interest of expediency. Getting back to acting with integrity, you must do what you say you're going to do.

Be Your Project Team's Biggest Fan

We discussed earlier that, as a project manager, you're the one responsible for clearing roadblocks for your team and not allowing them to be pressured by stakeholders or anyone else. The more you anticipate questions or issues from your stakeholders, the more you can clear the way for your team.

At the same time, you will want the team to do their best work. That could mean challenging them on occasion, or asking if they've thought of an alternative solution to a problem. Don't let them give up on a tough task or get discouraged when an ordeal seems to be more difficult than everyone thought at first. Celebrate the wins and continue to encourage and coach during the entire project.

Demonstrate your support of the team and the process. Doing so sets the tone for stakeholders, so they too can take pride in the effort of the team. Likewise, it also encourages your team as well. When they know you have their backs, they will be more likely to put in that extra effort that often results in breakthroughs or superior products.

Key Takeaways

You don't have to be an ordinary effective project manager, but, with some additional tweaks, you can become next-level in your performance.

- Acting with integrity is a crucial behavior that you need if you want to be hired by preferred organizations and work with great team members.

- The more red tape you can remove from the process, the better.

- Being organized in time and space allows you to maximize the resources available to you.

- Understanding what makes your team players tick will help you motivate and work with them.

- Make sure you set appropriate expectations that you can maintain and hold onto throughout the entire project.

- Your project team needs a big fan, and that fan is *you*.

In the next chapter, you will learn how to up your game consistently with continuous improvement.

[14] https://www.verywellmind.com/the-myers-briggs-type-indicator-2795583
[15] https://discinsights.com/disc-theory
[16] https://www.psychologytoday.com/us/basics/big-5-personality-traits

CHAPTER

9

ALWAYS

BE

IMPROVING

CONTINUOUS learning is critical for becoming an effective project manager and maintaining that effectiveness. Technology is continually changing, and there are always new developments and circumstances on the horizon. Successful PMs enjoy learning new things anyway, and it's good for the brain too. Being able to learn from each experience, then applying those lessons will only help you master the skills of project management.

Curiosity

When everything changes regularly, there will always be questions asked to illuminate how those changes affect your work. Get to know the company and the reasons behind the project. Also, understand what's going on in the broader world that affects your project and your company.

For example, who are the key players in the industry? What are the new technologies being developed? What are the pressures on the industry as a whole and the company in particular? Think about the supply chain for resources for your project. Who are the major suppliers? Who are their competitors and what pressures are they facing?

There's always more to know. As a project manager, you'll never get to a point where you can say to yourself: *I know everything there is to know about this project, company, and industry.* On the other hand, if you ever do, recognize you are missing something. Some new technology you never dreamed of may appear on the scene to automate or speed up a process that you're using. A new player might emerge, or another project manager in your company may bring a better project management tool or process.

Staying curious keeps you sharp. It ensures that you're always looking to see what's new and improved, which means you can make your project even better. Besides, your career will grow as well.

Lessons Learned

At the end of the project, or sprint/iteration in agile, make sure the team and stakeholders meet to discuss lessons learned and document them. Making sure the teachings don't get lost is crucial for implementing changes and truly improving later. Identifying what went wrong is only the first step; the only way to get *better* is to use those learnings in the future.

At one company I worked for, the process was called "Plus/EBIs." At the meeting, someone would get out the easel and paper and write down the lessons. To avoid shaming and blaming, each person who spoke would note a Plus, which was something that went well, before naming an EBI: "Even Better If." EBIs were things that went wrong and how to avoid them for the next time.

Does it seem a bit silly to ensure that a Plus gets written down? It's not, specifically for reasons of morale, that you will want to indicate the things that went right. These are also lessons that can be transferred to future projects. Maybe the team discovered a new way of doing something that had positive results, or it was a method that allowed everyone to work to their full potential.

In addition to the "don'ts," people need to know what to do. They need to know what does work instead of wasting time figuring it out. Remember that you're trying to maximize your resources on all projects—if you know something works well, why wouldn't you want to copy it and use it for the next one? Documenting that resource ensures it's not forgotten months from when you created it.

Cognitive Skills

This is the type of intelligence that's applied to the ability to plan, think critically, and solve problems. It also provides the foundation for being able to learn from experience and apply it to new situations.

Effective PMs must know how to analyze the data they have, even though it is likely incomplete, and make decisions based on that imperfect information. As you work on more projects, you'll end up with higher complexity that you must manage on behalf of your team and stakeholders.

Being able to handle increasing amounts of complexity doesn't necessarily mean you need to have any kind of college degree, as long as you can think analytically and be a leader. There are certificates that you can attain, such as the Project Management Professional (PMP) issued by the Project Management Institute and recognized around the world (PMI, n.d.)[17]. If you're using Six Sigma as your methodology, various institutions issue Green and Black belts according to your experience. I, myself, am a Green belt, but I don't work on many Six Sigma projects. Other certifications are available, such as Scrum Master, as well.

Some organizations may require their project managers to have or obtain one of these designations, but not all do. They're helpful when you're starting and don't have a significant track record for project sponsors to look at because they know you've at least been trained in the methodology. If you've been in the project management arena for some time and have a solid background in the methodology already, you may not need the certification.

Being a project manager is not about degrees and certifications; it's about your skills and capabilities. That is, how well you communicate and think through problems to arrive at a solution, along with your ability to manage a team and other resources. If you don't have a college degree, experience counts for a lot in this field. Sponsors are looking explicitly for results. A project manager who can plan, analyze, deliver, and, most importantly, learn from previous projects and apply the lessons, is a star.

Continuous Improvement

Have you heard that a shark has to keep swimming or it will die? Project managers must keep improving or their careers will die. Effective PMs are interested in self-development and want to continue learning and getting better at their jobs anyway, even if it didn't have a significant impact on their careers. Invariably, it does.

Building your proficiency as a project manager is not only about the technical skills of project management, learning about new performance measurement tools, or project management software. Beyond those mentioned, it encompasses leadership qualities, which can always be improved.

Look for ways that you can work on all these skills. Obviously, projects are a vital way to improve your craft! However, you should consider other opportunities as well. You might take a course, either online or at the college campus. Remember, you don't need an MBA. Community colleges offer classes on these topics, and many project management organizations have online courses you can take anytime.

One great thing about increasing your leadership capabilities is that every community has plenty of organizations and positions where you can work on them. Volunteering allows you to interact with all kinds of different people so that you can practice your communication abilities as well.

Mentors who can guide you in your career and answer questions about the field can be extremely helpful. If your human resources department offers mentoring, take them up on it! Otherwise, look for a role model whom you would like to emulate in your work. If they aren't close to you geographically, you can have an online relationship too. That person can be from a different background than project management. However, what's important is their attitude and maturity in skills that are important to you.

Some people don't like the word "mentor" because they believe it implies a lot of work and commitment on their part. Thus, you may not want to walk up to someone and ask them bluntly to be your mentor. However, making a *commitment* is critical if you really want to achieve anything valuable in your career and life.

Organizational Training Plan

When working for a mature firm that builds the value of project management into their budget, they will have training for every employee. This will even be so for workers who may not be working on projects. It's still essential for them to understand the value of it and how it adds to the goals. Not everyone will undergo the same training, of course. Project managers may need to know how to build an effective team or use advanced project management features, but other employees and executives won't.

In mature organizations, each role has its own training plan. The instruction is tailored to the individual and aligned with business objectives. No training for training's sake, but each employee will have the guidance or a training plan necessary for their particular contribution to the firm. That way, everyone involved has a foundation of knowledge that supports all programs and projects. Such a training program requires resources from the organization.

The training plan must have a budget and a schedule for every team member to be trained. Not all of the training is necessarily cost-related, because some aspects, such as coaching, typically don't require a line item on the budget. The team would include all the executives (including the C-suite), upper management, line management, and employees. The plan itself is aligned with the business plan and objectives, along with the projects and programs that will allow the company to achieve its goals.

Fortunately for all, we live in the 21st century, so there are a variety of training options to choose from. Online training, onsite training, study materials, coaching, and mentoring are just a few delivery mechanisms available. They range from one-off training up to a whole education system. The trainers may be contracted out or located in-house. In addition to the new employee training that everyone would receive, a mature organization also ensures access to continuous learning.

Last, but certainly not least, the firm's training plan must include metrics around performance and benefits or goals achieved. There is no point in maintaining a training plan that isn't fruitful. Maybe the organization measures improvement in project performance as a result of the training, or value generated. The metrics are refined in conjunction with an annual review of the plan to see if any changes need to be made.

If you're not working for a mature organization that lays a training plan out for its project managers specifically, check with your management to see if any of these training packages are available. As noted earlier, the experience is vital for a PM. Still, continuous improvement requires target training that helps you improve on technical and leadership project management skills, as well as business and broader knowledge. Ensure that you block out time on your schedule that will be dedicated to training.

What if your organization doesn't offer training? That doesn't mean you get to skip it. You'll need to do research on your own (or have a team member do it) to find a package that will improve your skills and that of your team players. You must dedicate time for all to get better at their project management capabilities.

If you want to bring project management to the next level in your organization, explain the reasons and benefits, then propose some options to your management. Most executives do not know what project management skills must be developed. However, you have this knowledge, and if you explain the benefits and how the firm will achieve better results than ever before, you can win their buy-in.

Measuring Training's Return On Investment (ROI)

With massive amounts spent on training, depending on the size of the organization, there must be a measurable ROI to justify the expense for profit-conscious entities. ROI modeling is a systematic approach to comparing the cost of training to its benefit, and choosing programs that result in the biggest bang for your buck.

It starts with the training plan, which must be tied directly to the firm's business plan. The blueprint specifies the goals of the training, which should link back to the business objectives, the investment or budget, and the anticipated benefits. It bridges the current skills of employees and managers to the vision, where the firm initiates and implements projects successfully.

How can you best indicate the value of the training? The easiest way to measure is to send out a survey to the participants afterward, but this is entirely subjective and doesn't give any meaningful feedback as to the benefits of the training. Final exams measure the individual employee's learning as a result of the training, which is at least objective. However, they still don't address the actual value or effects to the bottom line. They only measure the performance of the trainer, the quality of training materials, and how much the employees learned.

A concrete way to test the benefits is to put the training to work, where the value measurement is implemented and tracked for a few months after the training to ensure that improvement actually takes place. Typical measures include productivity, error rates, and velocity. This is the best way for companies to determine whether they're benefiting from the training plan.

Project Manager-Specific Training

Effective PMs balance business, technical, and leadership skills, so their training must encompass all three. As you know, training won't get you all the way to effectiveness, but it is a key component of continuous growth. As a project sponsor myself, I looked for the experience and track record—not just the degrees or certificates potential PMs had attained. I also made sure that I saw time dedicated to learning and growing.

I didn't want project managers who'd taken a training course or two and thought that was all they needed. To make my projects successful, I knew I needed those who continued to learn and wanted to get better, whether their organization supported that or not.

At the entry level—for those just starting—sponsors expect to see a base of knowledge of project management, as well as problem-solving skills. The training that project managers require at this level include:

- Fundamentals of project management, including methods and reporting.
- Problem-solving.
- Managing conflicts.

One step up from entry level is where the project manager will need more leadership skills because projects are a bit more complex, as well as a step up in technical competence, including an understanding of best practices in project management. At this stage, they should know more about the business as well. As you would expect, the training goes deeper now.

- Tools and best practices.
- Active listening.
- Developing relationships.
- Team building.
- Organizational Project Management (OPM) framework.
- Cognitive intelligence.
- Business knowledge.

PMs at the advanced level learn program management tools and techniques, where a program is a large project consisting of smaller interdependent projects. Their leadership and business proficiency is more advanced, and they are capable of leading global or virtual teams.

- Improving EQ (emotional intelligence).
- Facilitation, mentoring, and coaching.
- Negotiation.
- Empowering and setting direction.
- Project financials.
- Global projects with distributed teams.
- Performance management.
- Organizational Change Management.

On the next level of senior project managers, the PMs understand and leverage the OPM framework, including the details of the business plan and how programs and projects deliver value to the business.

- Business planning.
- Value management.
- Program management and performance.

Finally, at the strategic level, a PM is involved in creating the business plan and generating value across the entire portfolio of programs and projects.

- Strategic business planning.
- Managing the portfolio of programs and projects.
- Portfolio performance management.

Each level builds on the one below it, and the PM must master every level through training and experience before ascending to the next.

You can see that, as you grow through the stages of project management, your technical, leadership, and business capabilities develop in complexity as well. Training at each level needs to support the goals of that stage effectively for a project manager to improve.

Finding a project manager role when you have the training but not the experience can be difficult. To overcome this, you can ask your manager to lead a small project for the group, or be assigned to a larger project and assist that PM with some of the management tasks. Another way to build these proficiencies is to assist a nonprofit or volunteer group with taking charge of an event. You can also search for companies that are offering entry-level PM employment or freelance jobs and apply for them. You need to combine your training with practice to grow as a project manager.

Remember that projects typically grow in complexity as they do size-wise. Don't assume that just because you successfully led a small project that you're ready to tackle a much larger one. Build on your skills and take the lessons learned from each project for the next one, and, over time, you'll have the experience and knowledge necessary to excel at leading larger projects. It's better to have a track record of success that you develop one step at a time, rather than biting off a bigger project than you can chew and damaging your reputation as a result. Project management isn't going anywhere. In fact, it's expanding as more organizations become mature and see portfolios of projects as good ways to add to the bottom line. There will always be another project to work on, as long as you can show good results to those looking to hire PMs.

Key Takeaways

Being an effective project manager means continuous learning and improving. Situations are always changing, so staying on top of them is a challenge to accept. Training plans are vital for improving the team's project management skills, and each level has a more complex skill set that must be mastered.

- Curiosity may not be right for cats, but it's necessary for PMs who want to grow in their careers.

- At the end of each project or iteration, the team and stakeholders must gather to capture the lessons learned so they can be implemented for the next one.

- Project managers need to develop their cognitive intelligence as well.

- Effective project management relies on continuous improvement from the team members, especially the PM.

- Mature organizations already have a training plan in place for all employees with dedicated funds, but project managers in other firms must seek out opportunities for team training.

- As project managers grow in their careers, their knowledge and training requirements grow in complexity.

---

[17] https://www.pmi.org/certifications

FINAL WORDS

NOW that you've reached the end of this book, you should have a good grasp of what makes an effective project manager. You need to combine technical, leadership, and business skills to complete challenging projects successfully. As you develop through the different stages of being a PM, you'll need more advanced capabilities and training.

We talked about how to start as an effective project manager. There are different capabilities required, such as being able to communicate at all levels of business and prioritize issues for yourself and the team. Projects operate under the Triple Constraint of time, scope, and cost. These three are challenging to balance, but your job, as a project manager, is to try to keep them proportional to the objectives that the client has. You're the one who is responsible for the success or failure of the project, and the one who will be held accountable to a large extent. All the team members, stakeholders, tasks, and deliverables are puzzle pieces, with you being the one who fits them together and communicates the overall vision of the project.

The key elements of a project include a focus on value creation and the maintenance of scope, so you don't end up with the dreaded scope creep. Done is better than perfect, and you are the one driving the completion. That relies on your ability to plan and commit to the project properly. Unfortunately, none of us live in an ideal world, so you will also need to be able to manage risks appropriately. You may see some threats coming as you look out across the horizon, but others may be unexpected. You can't sit back and wait for them to happen; instead, be proactive in countering them.

Competency as a project manager is essential, and some of that will develop the more projects you manage. At the beginning of it, you'll find it helpful to discover what your client is an expert in. Valuing the client is critical for working well with them. Effective PMs have solid technical skills, and they also provide the vision that the team and stakeholders work towards. They set the direction for the team, but

142

they don't micromanage. Instead, they empower team members to get the work done in a way that leverages their strengths.

In addition to a team that works well together under the PM's direction, a successful project has buy-in from client management at all levels, from the senior executives on down to the line. These managers demonstrate their support and provide resources as requested. Mature organizations have project management built into their business plans, and the projects align with business objectives. Everyone within the firm understands what's in it for them and acts accordingly.

Not all firms are mature when it comes to project management, and many PMs will find that they need to do the work to show management specific benefits of the project, and help them understand how they can demonstrate support. Organizational culture varies from those who value project management to those with negative attitudes or with little knowledge about it. PMs may find themselves assisting with organizational change management (OCM) to further their projects and have a better chance of success.

When the project manager has a project sponsor who understands, values, and supports their efforts, they also have a much higher chance of success. The sponsor is the one who owns and will champion the project. Best practices for PMs include meeting regularly with their sponsor and sitting down at the onset of the project to divide up responsibilities and ensure nothing is omitted. Mature firms will have effective sponsors for their projects, but project managers working outside those firms may need to help develop their sponsors into being the right kind of champions for the project.

Although PMs empower their teams, they also need to understand the various tasks and get down in the trenches with their team members when necessary. Planning the project isn't enough. The project manager must also see to its execution and properly closing out of the project. They ensure a solid transition from the project team to operations, such that the receiving team is trained on the deliverable and understands the value of it; otherwise, the project might die on the vine. If operations never takes ownership, all questions come back to the project team, who aren't equipped to handle the ongoing process. Whether the project is successful or not, the PM should avoid playing the blame game and help the team overcome obstacles.

Having excellent technical skills will help you prevent some of those obstacles in the first place. You need to understand the different methodologies and the pros and cons of each, as well as ways to measure project performance. You'll step in when necessary, so you need some knowledge of the team members' capabilities, which they bring to bear on the project. You know you need to maximize your resources, and automating and backing up processes will help you deliver on schedule. In addition, to be most effective, you'll need to understand the business you're working in and its strengths and weaknesses, as well as potential opportunities and threats.

So far, so good. However, if you want to level up your game, you can go even farther. Always act with integrity, and you'll find hiring managers and team members who will want to work with you. Reduce red tape and be as organized as you can, so you can use your resources efficiently.

A key trait for effective PMs is the desire for continuous learning. You're always curious and trying to learn more about your craft, field, and company. At the end of a sprint or project, you would gather the stakeholders and team together to document the lessons learned, both positive and negative. Then, you will put that knowledge into action in future projects.

As an effective project manager, you would continue to obtain training for you and the team. Mature organizations have training plans and budgets, but you may need to do the research and develop one when you're not working for a mature firm. As you progress through the stages of being a project manager, from entry-level to strategic, your training will become more advanced. Although experience is important, training keeps you at the top of your game.

If you only take one idea away from this book, I hope it's the recognition that you can be effective without necessarily adding any degrees or certifications, and that you can get additional PM experience. Training does help you hone your natural talents and provides a structured method for improving your skills, yet you just need to start with the willingness to be teachable and a desire to grow and learn throughout your career.

Take what you've learned here and apply it, but also pass it on to your juniors. The more competent project managers the world has, the better. Now, you have the knowledge to go forth and be an even more effective project manager.

LEAVE A REVIEW

I would be incredibly *thankful* if you could take just 60 seconds to write a brief review on Amazon, even if it's just a few sentences.

If you have downloaded the bonus checklist for getting support from the managers, sponsor and stakeholders (the link is at the beginning of the book), you can attach it to your review and share your experience. This will inspire and encourage other fellow project managers who may have difficulties getting necessary support from their project partners.

Please log in to your Amazon account, then find this book *Being an Effective Project Manager*.

Alternatively, type this link into your browser, or scan the QR code:
www.amazon.com/dp/B08KQ2HT2X

Customer Reviews

★ ★ ★ ★ ★ 51

4.8 out of 5 stars ▼

5 star		94%
4 star		2%
3 star		0%
2 star		2%
1 star		2%

See all 51 customer reviews ▸

Share your thoughts with other customers

> Write a customer review

My Other Books You Will Love

For a complete collection of books, visit the author page

amazon.com/author/rsagile

rsagile.com

REFERENCES

Ali, J. (2017, December 21). **7 Tips on How to be a great Project Manager.** Retrieved from https://www.projecttimes.com/articles/7-tips-on-how-to-be-a-great-project-manager.html

Aston, B. (2020, April 29). **Essential Project Management Skills For 2020 (+How To Build Them).** Retrieved from https://thedigitalprojectmanager.com/project-management-skills/#soft-skills

Aziz, E. (2015, October 10). **Project Closing.** Retrieved from https://www.pmi.org/learning/library/importance-of-closing-process-group-9949

Barker, I. (2019, May 7). **97 percent of companies now use agile development methods.** Retrieved from https://betanews.com/2019/05/07/state-of-agile-report/

Bryce, T. (2006). **Why Does Project Management Fail?** Retrieved from https://www.projectsmart.co.uk/why-does-project-management-fail.php

Business Dictionary. (n.d.). **What is the Peter principle? definition and meaning.** Retrieved from http://www.businessdictionary.com/definition/Peter-principle.html

Cherry, K. (2019, July 17). **Myers-Briggs Type Indicator: The 16 Personality Types.** Retrieved from https://www.verywellmind.com/the-myers-briggs-type-indicator-2795583

CMOE. (2020, May 4). **5 Ways To Improve Your Strategic Thinking Skills Today.** Retrieved from https://cmoe.com/blog/improve-strategic-thinking-skills/

Colman, H. (2020, April 29). **Employee Training Metrics: How to Measure eLearning Effectiveness.** Retrieved from https://www.ispringsolutions.com/blog/training-metrics-how-to-measure-elearning-effectiveness

Curtis, L. (2019, October 26). **Five Must Have Core Competencies for Project Managers.** Retrieved from https://www.mpug.com/articles/five-must-have-core-competencies-for-project-managers/

DISC Insights. (n.d.). **DISC Theory and DISC Personality Traits.** Retrieved from https://discinsights.com/disc-theory

Florida Tech. (n.d.). **The Importance of Organizational Culture to Project Management.** Retrieved from https://www.floridatechonline.com/blog/business/the-importance-of-organizational-culture-to-project-management/

Gay, B. (2018, October 3). **What it Takes to be an Effective Project Manager.** Retrieved from https://www.projecttimes.com/articles/what-it-takes-to-be-an-effective-project-manager.html

Gilbert, J. (2012, March 27). **The Change Management Life Cycle; Involve Your People to Ensure Success.** Retrieved from https://www.batimes.com/articles/the-change-management-life-cycle-involve-your-people-to-ensure-success.html

Harned, B. (n.d.). **How to Become a Successful Project Manager.** Retrieved from https://www.teamgantt.com/guide-to-project-management

Harrin, E. (2018, December 17). **15 Essential Skills Every Project Manager Needs.** Retrieved from https://www.strategyex.co.uk/blog/pmoperspectives/15-skills-project-managers-will-need-2015/

Haughey, D. (2020, May 24). **Stop Scope Creep Running Away With Your Project.** Retrieved from https://www.projectsmart.co.uk/stop-scope-creep-running-away-with-your-project.php

Haughey, D. (2011, December 19). **Understanding the Project Management Triple Constraint.** Retrieved from https://www.projectsmart.co.uk/understanding-the-project-management-triple-constraint.php

HBS Online. (2017, November 2). **A 3-Step Change Management Framework for Businesses.** Retrieved from https://online.hbs.edu/blog/post/a-3-step-framework-for-managing-organizational-change

James, V. (2013, October 29). **Strategies for Project Sponsorship.** Retrieved from https://www.pmi.org/learning/library/strategies-project-sponsorship-5875

Kerzner, H. R. (2009). **Project Management Case Studies (3rd ed.).** New York, NY: Wiley.

KonMari. (n.d.). **About KonMari | The Official Website of Marie Kondo.** Retrieved from https://shop.konmari.com/pages/about

Leadem, R. (2018, August 12). **Why Emotional Intelligence Is Crucial for Success (Infographic).** Retrieved from https://www.entrepreneur.com/article/318187

Malsam, W. (2020, February 11). **What Is a Project Sponsor? Defining This PM Role.** Retrieved from https://www.projectmanager.com/blog/what-is-a-project-sponsor

May, A. (2016, December 14). **Stakeholder Buy-In: The Secret to Project Success.** Retrieved from https://www.dashe.com/blog/the-importance-of-stakeholder-buy-in

PM Tips. (2019, October 15). **Effective Project Leadership and Stress.** Retrieved from https://pmtips.net/article/effective-project-leadership-and-stress

Project Management Institute. (2018). **The Standard for Organizational Project Management.** Retrieved from https://www.pmi.org/pmbok-guide-standards/foundational/organizational-project-management

Psychology Today. (n.d.). **Big 5 Personality Traits.** Retrieved from https://www.psychologytoday.com/us/basics/big-5-personality-traits

Reichel, C. (2006). **Earned Value Management Systems (EVMS).** Retrieved from https://www.pmi.org/learning/library/earned-value-management-systems-analysis-8026

Santos, I. A. M. dos, Barriga, G. D. C., Jugend, D., & Cauchick-Miguel, P. A. (2019). **Organizational factors influencing project success: an assessment in the automotive industry, Production vol. 29.** Retrieved from https://doi.org/10.1590/0103-6513.20180108

Schibi, O., & Lee, C. (2015, October 10). **Project Sponsorship.** Retrieved from https://www.pmi.org/learning/library/importance-of-project-sponsorship-9946

Shenoy, A. (2014, January 29). **Bringing Vision to Your Projects: How to Excel as a Project Manager.** Retrieved from https://www.projecttimes.com/articles/bringing-vision-to-your-projects-how-to-excel-as-a-project-manager.html

Suda, L. V. (2007). **The Meaning and Importance of Culture for Project Success.** Retrieved from https://www.pmi.org/learning/library/meaning-importance-culture-project-success-7361

van Rooy, D. (2020, February 6). **7 Ways to Adopt a Proactive Mindset - and Achieve Success.** Retrieved from https://www.inc.com/david-van-rooy/7-ways-to-adopt-a-proactive-mindset.html

Wikipedia contributors. (2020, May 25). **Peter principle.** Retrieved from https://en.wikipedia.org/wiki/Peter_principle

Wikipedia contributors. (2020, April 1). **Seagull management.** Retrieved from https://en.wikipedia.org/wiki/Seagull_management

Windsor, G. (2020, March 3). **How to Work Effectively with Your Project Sponsor.** Retrieved from https://www.brightwork.com/blog/work-effectively-project-sponsor

Wrike. (n.d.). **What Is Portfolio in Project Management.** Retrieved from https://www.wrike.com/project-management-guide/faq/what-is-portfolio-in-project-management/

Wrike. (n.d.). **What is Program Management vs. Project Management?** Retrieved from https://www.wrike.com/project-management-guide/faq/what-is-program-management-vs-project-management/

Zilicus. (n.d.). **A Project Leadership Or Project Management - Becoming Effective Project Leader - Part II.** Retrieved from http://zilicus.com/Resources/blog-2014/Project-Leadership-Or-Project-Management-Becoming-Effective-Project-Leader-Part-ii.html

GLOSSARY AND ABBREVIATIONS

Agile/APM: Agile Project Management, methodology, or framework .... ch. 1
Agile Manifesto: 12 core values of APM defined in 2001 .... ch. 7
Analysis paralysis: Overthinking; not allowing decision making .... ch. 2
BCWP: Budgeted Cost of Work Performed (aka earned value) .... ch. 7
Black Belt/Master Black Belt: Six Sigma certification .... ch. 1
CCO: Chief Communications Officer .... ch. 4
CEO: Chief Executive Officer .... ch. 4
CPC: Cost Per Click (measurement of advertising expenses) .... ch. 4
DISC: Dominance, Influence, Steadiness, Compliant (personality types) .... ch. 8
EQ: Emotional Intelligence (understanding and managing emotions) .... ch. 1
Gantt Chart: A visual scheduling diagram named by Henry Gantt .... ch. 7
Global project: A project with members in different locations .... ch. 9
Jack of All Trades: Someone who is decent at everything .... ch. 1
Kanban: APM method for improving workflow; uses visual tools .... ch. 7
KPIs: Key Performance Indicators .... ch. 4
Leadership: A combination of skills, characteristics, and behaviors .... ch. 3
Marie Kondo: Wrote books about organizing, known as KonMari method .... ch. 1
MBTI: Myers-Briggs Type Indicator (personality types) .... ch. 8
Metrics: A system or set of measures, such as profitability, effectiveness .... ch. 4
MQ: Managerial Intelligence (management skills) .... ch. 1
OCEAN: Five personality type factors .... ch. 8
OPM: Organizational Project Management .... ch. 9
Organizational culture: employees' attitudes, behaviors, and values .... ch. 4
Plus/EBIs: Plus and Even Better Ifs (APM improvement practice) .... ch. 9
PM: Project Manager .... ch. 1
PMI: Project Management Institute .... ch. 9
PMP: Project Management Professional, certification .... ch. 9
Portfolio: Groups programs and projects .... ch. 4
Program: Consists of a group of related projects .... ch. 4
Project Manager: Leads the project and project team .... ch. 1
Risk Management: Foreseeing and addressing potential risks .... ch. 2
ROI: Return on Investment (return per dollar invested in the project) .... ch. 4
Scope Creep: Continuously adding new deliverables to the scope .... ch. 2
Scrum: The most popular APM method for developing complex products .... ch. 7
Scrum Master: A coach, promotes productivity, can be a project manager .... ch. 7
Six Sigma: A process improvement framework using statistics measures .... ch. 7
Sponsor: Has the most significant interest in the project's outcome .... ch. 5
Sprint: An iteration or cycle in the APM that lasts about a month .... ch. 7
Stakeholder: Has a share in the outcome of the project .... ch. 1
Triple Constraint triangle: Project limitations—cost, time, scope .... ch. 1
VP: Vice-President .... ch. 4
Waterfall: A traditional PM methodology; the chart resembles a waterfall .... ch. 1
WBS: Work Breakdown Structure (a tree structure with activities) .... ch. 7
WII-FM: What's In It For Me or Them (a way of communication) .... ch. 4
XP: Extreme Programming, an APM framework for software development .... ch. 7

BEING AN EFFECTIVE
PROJECT MANAGER

2020

CPSIA information can be obtained
at www.ICGtesting.com
Printed in the USA
BVHW030620091120
592835BV00022B/290